Theory's Practice: Reflections on Technology in Germany around 1800

Technology and Change in History

VOLUME 19

The titles published in this series are listed at *brill.com/tch*

Theory's Practice: Reflections on Technology in Germany around 1800

By

Jocelyn Holland

BRILL

LEIDEN | BOSTON

Cover illustration: Chisels from the workshop of Jim Eisenstein. 10 November 2024, Pasadena, California.

Library of Congress Cataloging-in-Publication Data

Names: Holland, Jocelyn, 1973- author
Title: Theory's practice : reflections on technology in Germany around 1800 / by Jocelyn Holland.
Description: Leiden ; Boston : Brill, 2025. | Series: Technology and change in history, 1385-920X ; volume 19 | Includes bibliographical references and index.
Identifiers: LCCN 2025011417 (print) | LCCN 2025011418 (ebook) | ISBN 9789004549531 hardback | ISBN 9789004549548 ebook
Subjects: LCSH: Technology—Germany—Philosophy—History—18th century | Technology—Germany—Philosophy—History—19th century
Classification: LCC T26.G3 H65 2025 (print) | LCC T26.G3 (ebook) | DDC 609.43/09033—dc23/eng/20250505
LC record available at https://lccn.loc.gov/2025011417
LC ebook record available at https://lccn.loc.gov/2025011418

Typeface for the Latin, Greek, and Cyrillic scripts: "Brill". See and download: brill.com/brill-typeface.

ISSN 1385-920X
ISBN 978-90-04-54953-1 (hardback)
ISBN 978-90-04-54954-8 (e-book)
DOI 10.1163/9789004549548

This book is printed on acid-free paper and produced in a sustainable manner.

Contents

PART 3
General vs. Historical Technology, circa 1800

Translator's Note

This translation wants to preserve a sense of the historical moment in which the texts were written. The result might, at times, be jarring to modern ears – such as when one comes across "fabricks" instead of "factories" – but there is an equal degree of pleasure to be found in the discovery of technique-specific words which have long since faded from common usage. Who knows today what it means to "calender," or how to use sugar scissors? This translation is also very much indebted to the exhaustive work of early nineteenth-century lexicographers such as John Ebers and Nathan Bailey, whose general bilingual dictionaries include a number of technical terms, as well as to writers of technical dictionaries, such as Alexander Tollhausen, who recorded the most specialized tools and procedures for a number of the arts, in multiple languages.

Introduction: the Past, Present, and Future of "Technology," circa 1800

> In a moment when the meaning of technics is indisputably one of the most essential questions facing our species, do we really want to make it an intellectual misdemeanor to ask big questions about "technology" and its historical role, however ill-defined the category is?
>
> JOHN DURHAM PETERS, "You Mean My Whole Fallacy Is Wrong"[1]

•••

> Also possible, but never advisable, is the further division of technology into the *theoretical* and *practical* part.
>
> GEORG LAMPRECHT, *The Instruction Book of Technology*[2]

∴

In his 2006 essay, "'Technik' Comes to America," Eric Schatzberg formulates the central problem of thinking and writing about technology as follows: "Anglo-American historians of technology have been remarkably incurious about *technology*, the most central term in the field."[3] It is a problem that others have identified as well. Guido Frison already commented in 1993 that "the present literature on the problems of *Technologie* and *Technik* is indeed lacking due to a common factor, i.e. the use of a non-critical, conventional concept of 'technology' which is almost always without a historical and sociological

1 John Durham Peters, "You Mean My Whole Fallacy Is Wrong" *Representations* 140 (Fall 2017): 13.

2 Lamprecht, "Introduction," *Lehrbuch der Technologie oder Anleitung zur Kenntniß der Handwerke, Fabriken und Manufacturen* [Instruction book of technology or guide to the knowledge of handcrafts, fabricks, and manufactures] (Hemmerdesche Buchhandlung, 1787). See pp. 116–147 of this edition.

3 Eric Schatzberg, "'Technik' comes to America: Changing Meanings of Technology Before 1930," *Technology and Culture* 47, no. 3 (July 2006): 486. Schatzberg includes a few exceptions to this generalization in the work of Ruth Oldenziel, Leo Marx, and Ron Kline.

 | DOI:10.1163/9789004549548_002

dimension."[4] Likewise, in an essay volume devoted to Greek thought and culture, François de Gandt cautions against the overuse of the term when he observes that "indications of a technical culture do not constitute a technology per se, that is, an organized discourse on tools and methods."[5]

Schatzberg's extensive writing about technology's conceptual history has done much to correct this lacuna by tracing technology's emergence as a concept against the backdrop of historical developments in theories of technical thinking. His *Technology: Critical History of a Concept* (2018) contains chapters devoted to the Greek *technê* and Latin *ars* as well as others on medieval strategies for dealing with the mechanical arts in systems of classification of knowledge, the rapid expansions of manufacture and industry that take place during and after the Enlightenment era, and the further developments that inform nineteenth- and twentieth-century thinking about technics and technology. The present study focuses on a particular moment in this longer history. Where the late eighteenth-century German context is concerned, these details are worth lingering over, because they reveal various attempts to reflect on technology's hybrid status as a science that incorporates theoretical speculations even as it remains one rooted in empirical objects of study. Throughout, I translate *Wissenschaft* as science, with the caveat that the late eighteenth-century idea of science differs from ours. Frans de Bruyn is one of many to comment that "terms that appear familiar and settled to us had different meanings in the eighteenth century," giving the example of science as "any subject explored systematically, with a view to uncovering its general or theoretical principles" to explain why cameralism, for example, was designated as a science and taught in the faculty of philosophy.[6] The same holds in the English context. Sara Landreth, for example, has commented on Samuel Johnson's broad definition of science as "not only 'Certainty grounded on demonstration' but also 'any art or species of knowledge,' including 'the seven liberal arts.'"[7]

Readers unfamiliar with those primary sources from the eighteenth and early nineteenth centuries which refer to technology might have the

---

4 Guido Frison, "Linnaeus, Beckmann, Marx and the Foundations of Technology. Between Natural and Social Sciences: a Hypothesis of an Ideal Type," *History and Technology* 10 (1993): 139.

5 François de Gandt, "Technology," in *Greek Thought: a Guide to Classical Knowledge*, ed. Jacques Brunschwig and Geofffrey E. R. Lloyd, with Pierre Pellegrin, trans. Selina Stewart and Jeannine Pucci (Harvard UP, 2000), 488.

6 See Frans de Bruyn, "Introduction," in *The Cambridge Companion to Eighteenth-century Thought*, ed. Frans de Bruyn (Cambridge University Press, 2023), 6.

7 See Sara Landreth, "Science in the Long Eighteenth Century," in *The Cambridge Companion to Eighteenth-century Thought*, 58. One of the reviewers of this manuscript provided a helpful reference from Daniel Breazeale's English translation of Fichte's *Doctrine of Science*.

impression that it is a concept of "marginal status" in multiple regards.[8] For example, Schatzberg mentions that "*technology* in any of its forms remained rare through the end of the eighteenth century" and that "Samuel Johnson's famous English dictionary of 1755 did not include it," nor did subsequent editions.[9] In Schatzberg's account, technology around 1800 is more of a "specialized term" with a "common set of meanings" in the German, French, and English contexts; "except in anomalous usage, [these meanings] did not refer to industrial processes or artifacts."[10] Elsewhere, Schatzberg defines *Technologie* in the early nineteenth-century German context primarily in terms that underscore its unimportance or deficits: as "a minor scholarly field in technical universities," it was "not the focus of any sustained theoretical discourse in the nineteenth century, not even in German."[11] There was, in plain speaking, "no *technology* before the twentieth century" because the term which existed "referred primarily to the 'science of the arts,' or more narrowly to treatises on the arts or descriptions of technical terminology."[12]

This depiction of technology's status at the turn of the nineteenth century needs more definition. For example, even if Samuel Johnson's *Dictionary of the English Language* did not contain an entry on technology, others did: the dictionaries of Elisha Coles (1676), Daniel Fleming (1763), John Ash (1775), and Nathan Bailey (1755) each include the term. The real question, however, is not how many times the word "technology" was used in various antiquated meanings, but rather what self-described "technologists" did with this concept in the years leading up to 1800 and afterward. Among these technologists are Johann Beckmann, Georg Lamprecht, and Johann Poppe, whose work comprises the focus of this book. For these three scholars (among others), technology was a matter of historical reflection, practical usage, and future speculation. As a closer look at the discursive practices that inform their writings will show, they grappled with fundamental questions regarding technology, the possibility of its categorization, and its economic and societal role. At stake is the very concept of technology itself. Thus, we should be willing to re-evaluate, with an open mind, how Beckmann and his contemporaries think about technical objects and create various categories for distinguishing between technical tools and processes. After all, as Hans-Peter Müller points out in an essay

8 Eric Schatzberg, *Technology. Critical History of a Concept* (University of Chicago Press, 2018), 3.

9 Schatzberg, *Technology*, 77. The first edition of Samuel Johnson's dictionary to include the term "technology" was published in 1818.

10 Schatzberg, "'Technik' Comes to America," 487.

11 Schatzberg, *Technology*, 74.

12 Schatzberg, *Technology*, 74.

on Johann Beckmann and Carl Linnaeus, "every classification … necessarily contains open or hidden theoretical reflections, because the systemization of knowledge is not possible otherwise."[13] And one can also include the moments of self-awareness that are common to Beckmann's, Lamprecht's, and Poppe's projects, as they reflect upon the difficulty of mediating between technical objects, technical practices, and the frameworks generated to adequately contain and organize them. However the history of technology from the twenty-first century perspective might have decided otherwise, an engagement with a complex idea of technology was absolutely central to the intellectual projects of writers such as these. The essays and translations contained in this volume have therefore all been undertaken with the purpose of shedding more light on a chapter in the history of technology which has often been overlooked, particularly in the Anglophone context.

This project has been constructed with two main objectives in mind: a theoretical one and a practical one. The practical objective is to make available the first English translations of a few key texts in German writing on technology around 1800. This is a service for a community of readers who might have arrived at the question of what technology means in the German context around 1800 from various disciplinary angles: scholars of history, philosophy, political science, economics, and literature who are invested in learning more about how the concept of technology functions in their particular field. The theoretical objective has more to do with filling in a blind spot where the conceptual history of technology is concerned than it does with correcting errors in existing scholarship, although it does ask that current generalizations about the status of the discourse on technology at the end of the eighteenth century be finetuned. The introductory essays found at the beginning of each section of this volume will do most of the legwork. They analyze the translations with an eye to the question of what insights these works have to offer us into a speculative thinking about technology, well before anything as coherent as a philosophy of technology (or technics) emerged on the scene toward the end of the nineteenth century. To give readers a preliminary sense of what to expect from reflections on technology around 1800, they generally do one or more of the following:

13 "Johann Beckmann und Carl von Linné – Anmerkungen zu einer prägenden Begegnung" [Johann Beckmann and Carl von Linné – comments on a formative encounter] in *Johann Beckmann (1739–1811). Beiträge zu Leben, Werk und Wirkung des Begründers der Allgemeinen Technologie* [Contributions to life, work and impact of the founder of the general technology], ed. Günter Bayerl and Jürgen Beckmann (Waxmann, 1999), 96.

- use historical narratives as a springboard for more general thinking about technology
- distinguish between various "kinds" of technology around 1800
- experiment with different kinds of technical order and critique the ideas of order that govern the systems of technology
- think about the relation between technical objects and technical practices
- think about technology in terms of various kinds of mobility (an idea that has a long historical trajectory)

To be clear: such passages of theoretical reflection do not exist in a vacuum. For every moment of speculation about technology, there is an abundance – often, an excess – of historical, bibliographic, and personal details.

1 Theoretical Perspectives: Carl Mitcham and Keith Tribe

The inspiration for this project comes from a historical side angle that might seem counterintuitive: German Romanticism's interest in technological thinking and concepts. Although better known for their thinking about organic life and generation, writers such as Friedrich von Hardenberg (Novalis) received technical training in mining and other fields and lived two lives, as literary thinkers and government administrators. Emblematic of this dual status is the fact that Novalis had Beckmann's guide to technology in his library – his was a familiar name. One of the things that characterizes German Romantic writing, especially in its early years, prior to 1800, is a heightened sensitivity to the nuances of language. And while it might seem like an impractical exercise to attune oneself to such nuances in texts like the preface or introduction to Beckmann's *Guide to Technology*, part of the justification for doing so comes from the historian Keith Tribe, who has taken a comparable approach to the concept of economy. In the opening argument to *The Economy of the Word: Language, History, and Economics*, Tribe states that "the use of economic language is the proper object of the history of economic discourse."[14] Such a focus on language calls for, as he describes it, "particular instruments and techniques" which "derive broadly from the history of the sciences, literary analysis, linguistics, and philology."[15] In the context of his study, the particular idea of an "economy" of the word is meant to refer to a "complex and ever-changing

14 Keith Tribe, *The Economy of the Word. Language, History, and Economics* (Oxford University Press, 2015), 2.

15 Tribe, *The Economy of the Word*, 3.

semantic system."[16] If one were to direct this same focus toward a cross-section of technological writings around 1800 then, arguably, the reward would be a better understanding of the complexity of technology as a concept than one is able to achieve either through more far-reaching historical accounts, where writers like Beckmann appear as a footnote, or through approaches less reliant on extensive textual engagement.[17]

The work of the philosopher Carl Mitcham has also helped shape the present study. In his *Thinking through Technology*, Mitcham makes a broad distinction between an "engineering philosophy of technology" and a "humanities philosophy of technology." The former takes place through "analyses of technology from within" and is "oriented toward an understanding of the technological way of being-in-the world as paradigmatic for other kinds of thought and action."[18] Humanities philosophy of technology, by contrast, is "the attempt of religion, poetry, and philosophy to bring non- or trans-technological perspectives to bear on interpreting the meaning of technology."[19] Both of these interpretive frameworks are useful for a reading of technological writing around 1800. The engineering philosophy of technology, for example, offers a reference point for how to understand theories and acts of "translation," an idea developed by Francis Bacon and clearly thematized in the German technological writing. Techniques connected to certain actions – dividing, binding, etc. – are the currency for the transfer and application of technical actions in new contexts. The humanities philosophy of technology is relevant as well, when one considers the broad discursive reach of technological writing around 1800. It draws directly from French materialist philosophy, in the form of Diderot's and D'Alembert's work on their encyclopedia project, and also makes frequent contact with the German literary tradition.

Mitcham's thoughts about the difference between Greek *technê* and modern technology have also been useful when trying to understand attempts by

---

16 Tribe, *The Economy of the Word*, 2.

17 There are, of course, exceptions, and the work of Ulrich Troitzsch is foundational in this regard. Troitzsch was one of the first to recognize that Beckmann's technology concept spans both the "doctrine of acquiring and processing raw materials" as well as the "thoughtful application of technics [*Technik*] for the economy." Ulrich Troitzsch, *Ansätze technologischen Denkens bei den Kameralisten des 17. Und 18. Jahrhunderts* [Approaches to technological thinking by the cameralists of the 17th and 18th centuries] (Duncker und Humboldt, 1966), 8. Troitzsch's research on Beckmann has traced the "beginnings of technological thinking" with an eye to the broader connection of economy and technics as opposed to a strict history of technological progress (Troitzsch, *Ansätze*, 8).

18 Carl Mitcham, *Thinking through Technology. The Path Between Engineering and Philosophy* (University of Chicago Press, 1994), 39.

19 Mitcham, *Thinking Through Technology*, 39.

German technologists to allow particular technical practices to be the building blocks of their systems. He writes, for example, that whereas there is no *logos* that directs "the actual process of production" for the activities of *technê* in the Greek context, this is by contrast "precisely what modern technology proposes to furnish – a *logos* of the activity, a rationalization of the process of production, independent of, if not actually divorced from, any particular concept of *eidos* or form."[20] On the basis of such a distinction, one can read the classifications of the verbs (as "processes") of technical work that one finds in Beckmann's general technology in a more emphatically modern light.[21] It also shows how the close analysis of historical changes in semantics is indispensable for understanding how technology in the modern sense was able to emerge and evolve as a concept, and ultimately as a self-perpetuating force of continued technological innovation.

Some historians of technology, like Andrew Jamison, have made a distinction between different kinds of early "theorists of technological innovation."[22] For example, Jamison places Adam Smith (as someone with "little interest ... in the actual process of technical production, for whom "technology was considered as a producer of wealth") in a very different camp than Beckmann (who was "interested in studying technical production as a science, in drawing out its general laws or principles").[23] Without subscribing to a wholesale division between these two lines of thought, the present study is at the same time very interested in placing more attention on the idea that the pursuit of "general laws and principles" where technology is concerned is a strong motivation for Beckmann.

2 Narratives of *technê* and Categories of Knowledge

German discussions of *Technologie* around 1800 are positioned at the confluence of multiple histories which intersect and disconnect over time, including the history of craft knowledge associated with the Greek *technê* and Latin *ars*,

20 Mitcham, *Thinking through Technology*, 128.

21 This phenomenon corresponds to what Antoine Picon has described as a "transition from a static understanding of structures to a more dynamic understanding of operations and functions." Antoine Picon, "Towards a History of Technological Thought," in *Technological Change: Methods and Themes in the History of Technology*, ed. Robert Fox (Routledge, 1996), 45.

22 Andrew Jamison "Technology's Theorists: Conceptions of Innovation in Relation to Science and Technology Policy," *Technology and Culture* 30, no. 3 (July 1989): 512.

23 Jamison, "Technology's Theorists," 512.

medieval and early modern approaches to categorizing knowledge according to shifting hierarchies of values, and the history of trade organizations, which included both guilds and various kinds of societies. (The Hamburg Society for the Improvement of the Arts and Useful Trades, for example, was founded in 1765 and saw itself tasked to improve the community and generally "realize the goals of the Enlightenment in all walks of life").[24] These individual histories take place within overarching narratives of technical achievement which have been plagued by ruptures, such as the "profound discontinuity … between the technological achievement of the Greco-Roman period and that of the Latin Middle Ages" that George Ovitt has observed,[25] and they also exist within a broad field of terms whose meanings are ever-changing due to time and translation. Phrases such as "practical arts" and "useful arts," like "technology" itself, have intricate discursive trajectories in the European linguistic context whose complexity can only be touched upon here. For example, the term "artes practicae" was already present in Medieval Latin, but the expression "practical arts" only became more common in late seventeenth-century English. One finds it, for example, in the subtitle "Mathematical and Practical Arts" of the *Mariners' Magazine*,[26] where the practical arts include various aspects of sea-faring and navigation. The German term *praktische Künste*, by contrast, only became commonplace in the eighteenth century. This brief historical overview seeks to identify various problem areas that were encountered along the way by those who attempted to discuss and define particular aspects of technical language and technical knowledge well before the concept of "technology" arrived on the scene.

In the context of the Greek and Roman philosophical tradition, there are ongoing discussions among scholars as to how to understand *technê* and *ars*, and how the knowledge of handcrafts relates to other types of knowledge. Those whose research is grounded in Greek or Roman classical literature and philosophy can also offer perspectives that differ significantly from those whose interests in these terms are rooted in the early history of medieval philosophy and craftsmanship. One of the key discussion points has to do not only with what *technê* is, but also what, if anything, makes it different from other types of knowledge, such as *episteme*, which is often considered more

24 Holger Böning, "Reimarus, Johann Albrecht Heinrich, 1729–1814," in *The Bloomsbury Dictionary of Eighteenth-century German Philosophers*, ed. Heiner F. Klemme and Manfred Kuehn (Bloomsbury, 2010), 615.

25 George Ovitt, "Technology and Science," in *The Cambridge History of Science*, vol. 2, Medieval Science, ed. David C. Lindberg and Michael H. Shank (Cambridge University Press, 2013), 633.

26 Edited by Captain Samuel Sturmy (1633–1669); printed in London for E. Cotes (1669).

intellectually stable or certain. In a work on the role of *technê* in Aristotle's ethics, the philosopher Tom Angier states that the *technê/episteme* distinction is not as sharp as people usually take it. Referring to the pre Platonic context, he comments that "in much of earlier Greek writing, *episteme* and *technê* are used interchangeably."[27] Angier is also cautious when it comes to the work of Plato and Aristotle, due to the fact that more modern conventions of distinguishing between the pure and applied sciences have often been retroactively applied to create a distinction between *episteme* as theoretical knowledge and *technê* as knowledge related to craftsmanship.[28] Serafina Cuomo has also suggested that the spectrum of *technê*'s meanings cannot be easily mapped onto modern concepts of knowledge, claiming that "essentially the same thing appears to have been variously called *technê*, *sophia* (knowledge/wisdom), *episteme* (rigorous/stable knowledge), or *dynamis* (power).[29]

Some scholars do, in fact, wish to differentiate more clearly between *technê* and *episteme*, and they do so in various ways. For Vincent Leitch, who speaks from a postmodern perspective, it is a question of dynamic as opposed to static kinds of knowledge: he emphasizes the "stability of *episteme*" as opposed to the "movement/dynamism of techne."[30] Others, such as Jacques Brunschwig, arrive at the same conclusion from a different methodological trajectory; he argues that "transforming opinions into knowledge is a move from instability to stability; *episteme*, through its root, evokes the ideas of cessation, rest, stability."[31]

Mitcham takes a somewhat softer stance. He sees "*techne* and *episteme*, art and systematic or scientific knowledge," as "closely associated" in Plato's work, while emphasizing that "techne (in Plato) is also conceived to be distinct from what we would call pure theory, or any knowledge that does not bear upon the material world in some practical manner."[32] Like many scholars, Mitcham

---

27 Tom Angier, Technē *in Aristotle's Ethics: Crafting the Moral Life* (Continuum, 2012), 2.

28 Angier, Technē *in Aristotle's Ethics*, 2.

29 Serafina Cuomo. *Technology and Culture in Greek and Roman Antiquity* (Cambridge University Press, 2007), 9. Luc Brisson likewise discusses how *sophia* could refer equally well to the knowledge possessed by carpenters, doctors, poets, etc. Luc Brisson, "Myth and Knowledge," in *Greek Thought: a Guide to Classical Knowledge*, ed. Jacques Brunschwig, Geofffrey E. R. Lloyd, Pierre Pellegrin, trans. Catherine Porter (Harvard University Press, 2000), 39.

30 Vincent B. Leitch, "Postmodern Theory of Technology: Agendas," in *symplokē* 12, no. 1/2 (2004): 212.

31 Jacques Brunschwig, "Epistemology," in *Greek Thought: a Guide to Classical Knowledge*, ed. Jacques Brunschwig, Geofffrey E. R. Lloyd, Pierre Pellegrin, trans. Catherine Porter (Harvard University Press, 2000), 86.

32 Mitcham, *Thinking through Technology*, 118–9.

turns to Aristotle's *Nicomachean Ethics* in order to ground his discussion of *technê*. In this work, Aristotle touches upon a broad number of topics connected to the question of how best to live, ranging from excellence of character to questions of virtue, friendship, and happiness. Mitcham's reading centers on Aristotle's definition of *technê* as a "habit (or stable disposition to act in a specific manner) with a true *logos* concerned with (or ordered toward) making (the human production of material objects)."[33] In Mitcham's interpretation, "*techne* is an ability to make that depends on correct awareness of or reasoning about the thing to be made. The absence of *technê* in an ability to make involves either the absence of any *logos* … or the presence of a false *logos*."[34] The "most fundamental difference" that separates Greek *technê* from modern technology, as Mitcham sees it, is that *technê* "involves *logos*, but only in grasping form, not in directing the actual process of production, the activity qua activity. There is no *logos* of this activity."[35]

There is general consensus that the *technê* concept, connected more or less loosely to different kinds of knowledge in Plato's dialogues, undergoes a gradual process of specification and hierarchical distinction in Aristotle's ethics. At the same time, the question of *technê*'s moral value has remained a matter of dispute, as it offers a possible way in which the structures of thought associated with practical knowledge can have important roles to play beyond artisanal contexts. Thomas Johansen, who has written on the concept of *technê* in the context of productive knowledge in ancient Greek philosophy, suggests that in moral and political thinking, "already prior to Plato, *technê* was used as a model for ethical and political knowledge."[36] Johansen contrasts arguments which claim that "practical knowledge, know-how, could not be reduced to propositional knowledge, knowledge-*that*," with the question of whether "knowledge-*how* may be understood propositionally along the lines of knowing that something is a way to do something."[37] For Rachel Barney, the answer is not clear cut. Even if, in Plato's early dialogues, "Socrates is obsessed with the idea that [virtue] might be [a craft] … We never see Socrates argue that virtue is wholly a matter of rational rather than affective factors, or that learning a craft requires only instruction. We do not see ancient philosophers citing various kinds of craft education to argue for different theories of moral

33 Mitcham, *Thinking through Technology*, 120.
34 Mitcham, *Thinking through Technology*, 120.
35 Mitcham, *Thinking through Technology*, 128.
36 Thomas Kjeller Johansen, "Introduction," in *Productive Knowledge in Ancient Philosophy. The Concept of* Technê (Cambridge University Press, 2021), 11.
37 Johansen, "Introduction," 9.

education, or arguing about just *which* craft provides the closest analogue to moral knowledge."[38] Rather, what Barney finds is that "the *techne* model is an exercise in what Aristotle calls reasoning from what is more familiar, or better known to us, to what is less so," with the idea that the same "care and effort" can be put into virtue that a craftsman would put into a menial task.[39] Tom Angier, who has also studied the role of *technê* in Aristotle's ethics, is somewhat more optimistic when disputing the argument that *phronesis* (Aristotle's concept of "full virtue") "cannot be a *techne*, because whereas the *technai* are directed at particular, circumscribed goods … human virtue is directed at 'the good life in general.'"[40] Instead, his thought is that this argument doesn't take into consideration whether human virtue could "be embodied in a *techne*": even if "ordinary linguistic usage may not make reference to such a highly general, all-inclusive craft," Plato nonetheless "spent much effort on exploring the possibilities for one, and it is not clearly objectionable to suggest one of these might be realized."[41] After all, if Aristotle has the concept of a "master-art," then "why not a 'master-art' of virtue"?[42]

Clearly, *technê* presents a troubling problem, as one grapples with what kind of knowledge is at stake and how to value it. That said, the precise "trouble with *techne*" according to Serafina Cuomo is of a different nature altogether. This trouble has to do with a "moral ambiguity" – that *technê* "does not necessarily provide knowledge about how to use the product."[43] At the heart of this problem is an idea that will, in various guises, follow discussions of *technê* for centuries to come: that of mobility. It has already been mentioned how the adaptability of *technê* was contrasted against the fixity of *episteme*. Cuomo

38 Rachel Barney, "*Technê* As a Model for Virtue in Plato," in *Productive Knowledge in Ancient Philosophy*, ed. Thomas Kjeller Johansen (Cambridge University Press, 2021), 63–5.

39 Barney, "*Technê* as a Model," 85.

40 Angier, Technē *in Aristotle's Ethics*, 41.

41 Angier, Technē *in Aristotle's Ethics*, 42.

42 Angier, Technē *in Aristotle's Ethics*, 42. Note that there has been strong resistance to this idea: "Ultimately, Socrates shows that technical knowledge is inadequate to venture into moral and political territory, and that, rather than a super-*techne*, an altogether different form of knowledge is required" (Cuomo, *Technology and Culture*, 29). Likewise, David Roochnik is against the idea that Plato "conceives of human life as amenable to technical treatment and therefore as rationally manageable," concluding that "Plato rejects *techne* as a model for moral knowledge." See David Roochnik, *Of Art and Wisdom. Plato's Understanding of* Techne (Penn State University Press, 1996), 6. Schatzberg's perspective is also relevant here, given that he states that "Aristotle's' distinction between *techne* and *phronesis* is just as categorical as his separation of *techne* and *episteme*" (Schatzberg, *Technology*, 21).

43 Cuomo, *Technology and Culture*, 39.

adds her own perspective on *technê*'s mobility when she describes two interconnected ways having to do with technical knowledge and technical practice. Technical knowledge, she suggests,

> is often described as conjectural or stochastic – literally, able to hit the target. The target in question is identified with the right moment or opportunity (*kairos*), which cannot be defined in advance but must be sought constantly, and varies from case to case. The stochastic character of *techne* is then the result of its constant negotiation between general principles and individual situations.[44]

Technicians themselves were also "associated with unregulated mobility and change."[45] For Cuomo, this is due to the fact that every *technê* "precipitates a dramatic change" by creating something that previously did not exist.[46] There is a corresponding change in the technician: "*techne*, as something that can be taught, changes people – it transforms laymen into experts" and "produces mutants, people born as one thing and transmogrified into another."[47]

Closely connected to the question of what the moral value of *technê* may or may not be is the social status of craftsmen. Alison Burford has described, with reference to Xenophon, Plato, and Aristotle, several reasons as to why "banausic" labor was held in disdain in Ancient Greece. These include a "softening of the body" that induces a comparable "softening of the mind,"[48] an inability to "practice virtue,"[49] and a lack of strong connection to the land such as was valued in the agricultural work of farmers.[50] Another factor, which Schatzberg has also pointed out, is the Aristotelian distinction that "ethics and philosophy were inherently virtuous, while the practical arts acquired virtue only by serving ends external to themselves."[51]

Longstanding perceptions regarding the value differences of various forms of knowledge contributed to medieval classifications. Pamela O. Long writes that the Aristotelian tradition "distinguished between episteme (theoretical knowledge of the unchanging); praxis (knowledge of contingent things requiring

44 Cuomo, *Technology and Culture*, 18.
45 Cuomo, *Technology and Culture*, 37.
46 Cuomo, *Technology and Culture*, 37.
47 Cuomo, *Technology and Culture*, 37.
48 Alison Burford, *Craftsmen in Greek and Roman Society* (Cornell University Press, 1972), 12.
49 Burford, *Craftsmen*, 34.
50 Burford, *Craftsmen*, 29.
51 Schatzberg, *Technology*, 4.

judgement); and *technê* (making and thinking about making things)."[52] These distinctions, in her historical retelling, feed into the epistemological hierarchies of the Middle Ages: "From the time of Aristotle, such categories were hierarchically ranked: episteme was at the highest level and technology at the lowest."[53] In this perspective, mechanical arts were kept beneath, and separate from, the liberal arts.[54]

3 The Mechanical Arts in Medieval and Early Modern Classificatory Schemes

Perceptions of the mechanical arts at the end of the eighteenth century – the point at which, according to Celina Fox, "the drive to describe and rationalize the mechanical arts … reached its peak"[55] – owe much to the visions and revisions of what position they should occupy in medieval and early modern classificatory schemes. Without going deeply into the rapidly shifting political and cultural landscapes of the territories that would eventually become central Europe, it is possible to outline some of the key developments of how the notion of a mechanical arts emerged and gradually began to assume the meaning familiar to us today.

The path of the Latin term *ars* – a translation of the Greek *technê* – is difficult to follow in its peregrinations through the Middle Ages.[56] The time period between Greek and Roman antiquity and the Latin Middle ages is marked by upheaval and transformation or, as George Ovitt refers to it, a "profound

52 Pamela O. Long, *Artisan/Practitioners and the Rise of the New Sciences, 1400–1600* (Oregon State University Press, 2011), 127.

53 Long, *Artisan/Practitioners*, 127.

54 George Ovitt, in his work on labor and technology in medieval culture, acknowledges the Aristotelian distinction between *episteme* as philosophic and scientific knowledge and *technê* as the productive arts, noting that, in the Greek view, "speculation was higher than practice" See *The Restoration of Perfection. Labor and Technology in Medieval Culture* (Rutgers University Press. 1987), 108–9.

Although *technê* and *episteme* were "fundamentally distinct categories" for Aristotle, Ovitt acknowledges that Plato "did little to distinguish techne from other forms of knowledge" and that his early dialogues "used techne and episteme (scientific knowledge) interchangeably, using both in the sense of expertise" (Ovitt, *Restoration of Perfection*, 20).

55 Celina Fox, *The Arts of Industry in the Age of Enlightenment* (Yale University Press, 2009), 16.

56 As Eric Schatzberg has noted, "we still lack a comprehensive history of the concept of *ars* and its counterparts in the medieval and early modern periods" (Schatzberg, *Technology*, 29).

discontinuity," such that "early-medieval Europe lost ground in many areas of technological development between the sixth and the ninth centuries."[57] At the same time, some areas of technological development gained ground, particularly in the areas of agriculture, water power, and textile production.[58]

In recent years, much attention has been focused on the social context of the mechanical arts, both in university and artisanal contexts.[59] In tandem with practical developments, a scholarly interest in the kind of knowledge required for various kinds of mechanical practices gradually developed during the centuries after Aristotle. Ovitt and others have pointed out that the "first fully developed classification and analysis of the mechanical arts was not contributed to the tradition of classification until the twelfth century."[60] Progress in this area was slow: earlier classificatory schemes, such as Varro's lost *Disciplinarum* [On the Disciplines] (116–127 BCE), or Martianus Capella's *De nuptiis philologiae et mercurii* [On the Marriage of Philology and Mercury] (early 5th century CE), did not make reference to "mechanical arts" per se, and between the earlier and later work, "lesser arts" such as medicine and architecture were removed from consideration.[61] The importance of such systems of classification cannot, however, be overstated. In Ovitt's words, "a classificatory scheme allows a culture to fit information into a preexistent order of educational, social, political, and economic assumptions. Furthermore, taken as a whole, a classificatory system allows both the system makers and its users to discover larger relationships among bodies of information."[62] Accordingly, the classificatory schemes of the Middle Ages were developed by theologians "to show the relation of human knowledge to the knowledge of God."[63]

Scholars of medieval culture generally agree that the *didascalia*[64] of the early twelfth century were the first schemes of knowledge to include the mechanical

57 Ovitt, "Technology and Science," 633–634.

58 Ovitt, "Technology and Science," 635f.

59 The term "artisan" changes according to historical and cultural contexts. James R. Farr has shown that the "deceptively simple question" of what an artisan was (he surveys the years 1300–1914) could be answered in such varying ways as "a guild member," a "skilled person," "a wage earner," and a someone who is essentially indistinguishable from a merchant. James R. Farr, *Artisans in Europe, 1300–1914* (Cambridge University Press, 2001), 3.

60 Ovitt, *The Restoration of Perfection*, 111.

61 Ovitt, *The Restoration of Perfection*, 111–112.

62 Ovitt, *Restoration of Perfection*, 108.

63 Ovitt, *Restoration of Perfection*, 108.

64 Jerome Taylor, "Introduction," in *The Didascalicon of Hugh of St. Victor*, trans. Jerome Taylor (Columbia University Press, 1961), 3. "Didascalic, or didactic, literature" was "concerned in various ways with what arts or disciplines a man should study and why he should acquire them" (Taylor, "Introduction," 3).

arts, of which the *Didascalicon* written by Hugh of St. Victor is the best known. The structure of this work, in which Hugh of St. Victor "aims to select and define all the areas of knowledge important to man" and demonstrate that these areas are "essentially integrated among themselves" as well as necessary for man's attainment of "human perfection" and "divine destiny," anticipates the encyclopedic efforts of later centuries, while remaining fully within a framework of Christian practices.[65]

In the *Didascalicon*, Hugh of St. Victor breaks the mechanical arts down into seven categories: "fabric making, armament, commerce, agriculture, hunting, medicine, and theatrics."[66] He groups them into two categories, such that three "pertain to external cover for nature" and "four to internal"; this distinction, in turn, aligns with the trivium and quadrivium: "for the trivium is concerned with words, which are external things, and the quadrivium with concepts, which are internally conceived."[67] To justify the designation of the sciences as mechanical, Hugh of St. Victor further explains that the term refers to the fact that "their concern is with the artificer's product, which borrows its form from nature" (Hugh of St. Victor, *Didascalicon*, 75). He also reinforces the distinction between the mechanical arts and the liberal arts on the basis of the status of those who originally practiced them in antiquity ("the populace and sons of men not free" for the former, "free and noble men" for the latter).[68] The chapter ends, however, with an all-encompassing perspective that allows mechanics, like other aspect of philosophy, to be understood in theoretical terms while at the same time invoking the physical labor of the human hand:

> In all this appears the great diligence of the ancients, who would leave nothing untried, but brought all things under definite rules and precepts.

65 Taylor, "Introduction" 3.

66 Hugh of St. Victor, *Didascalicon of Hugh of St. Victor*, 74.

67 Hugh of St. Victor, *Didascalicon of Hugh of St. Victor*, 74–5. Joan Cadden has also commented on the ways in which the mechanical arts, frequently numbered as seven, were "close but always in the shadow of the liberal arts." See Joan Cadden, "The Organization of Knowledge: Disciplines and Practices," in *The Cambridge History of Science*, vol. 2. Medieval Science, ed. David C. Lindberg and Michael H. Shank (Cambridge University Press, 2013), 245. She suggests that the mechanical arts, through this shadow relationship, served predominantly two roles: a negative one ("because they engage the body as well as the mind" and their subject is "merely human works"), and a positive one, as a "supplement to the liberal arts, particularly with respect to their engagement with the natural world" (Cadden, "Organization of Knowledge," 245).

68 Cadden, "Organization of Knowledge," 245.

> And mechanics is that science to which they declare the manufacture of all articles to belong.[69]

The foothold Hugh of St. Victor established for the mechanical arts in his *Didascalicon* was broadened by the late Middle Ages to include "crafts, building construction, and material processes of all kinds," as Pamela O. Long has shown in her extensive work on the status of artisans and crafts of the time period.[70] Even as she acknowledges substantial disagreements about how to define, identify, and describe the causes of "technological innovation" in the European context,[71] Long argues in favor of a "contextual view of the history of technology," one that

> does not look for modernity, or for signs of "progress" or "technological creativity" conceptualized from our own culture's point of view; nor does it look for causes of "Western superiority" as several economic historians have recently done. Rather, a contextual methodology attempts to understand technologies both customary and inventive within the ambience of the historical cultures in which they are found, and to understand their meaning for the diverse people who lived within those cultures.[72]

Setting aside for the moment the methodological question that arises when using the word "technology" to refer to time periods before the concept of technology existed – a question to which I will return at a later point in the introduction – Long's contextual approach is appealing for its willingness to focus on the complexities of particular historical moments without immediately relegating them to a historical long view. This same approach is also operative in the metaphor of "trading zones"[73] that Long uses to refer to the transmission of artisanal knowledge in fifteenth- and sixteenth-century Europe. As she understands it, these are arenas in which university scholars and craftsmen could intermingle: in which "the learned taught the skilled, and the skilled taught the learned, and in which the knowledge involved in

---

69 Hugh of St. Victor, *Didascalicon of Hugh of St. Victor*, 75.

70 Long, *Technology, Society, and Culture*, 1.

71 Long refers specifically to the work of M. M. Postan and Emmanuel Le Roy Ladurie, who have focused on population expansion and the inadequacy of traditional agricultural practices as well as to Robert Brenner, who has focused on economic crises in terms of class struggles (Long, *Technology, Society, and Culture*, 2).

72 Long, *Technology, Society, and Culture*, 5.

73 Long acknowledges that this term is borrowed from Peter Galison, who uses it in a very different historical context.

each arena was valued by both kinds of 'traders.'"[74] Such trading zones also facilitated a re-evaluation of the status of the mechanical arts, since it was required that "learned individuals valued practical and technical knowledge, not only for what it could achieve in the material world … but also as a form of knowledge," even as artisan/practitioners learned to value the classical textual tradition.[75] One could take Guidobaldo del Monte as an example of a "learned individual" who expresses his appreciation for the mechanical arts. Paolo Rossi has described Guidobaldo's perspective that "utility and nobility work jointly to adorn the mechanical disciplines" and that "[i]n mechanics, geometry achieves its complete fulfillment, and through mechanics, man achieves mastery over physical and natural things."[76] The Aristotelian view of the inferiority of human craftmanship, long regarded as a pale imitation of natural processes, was, in Elspeth Whitney's words, "gradually eroded over the course of the Middle Ages,"[77] engendering a "new alliance between *techne* and praxis," one which Schatzberg observes already in the fifteenth century.[78]

The seventeenth century heralded the arrival of a vocal proponent of the practice of the mechanical arts and the knowledge of their history. In *The Advancement of Learning* (1605), Francis Bacon outlines several reasons for valuing the mechanical arts. In addition to furthering the study of natural philosophy,[79] they also have a fundamental role to play in the "endowment, and benefit of man's life":

> But if my judgement be of any weight, the use of history mechanical is of all others the most radical and fundamental towards natural philosophy; such natural philosophy as shall not vanish in the fume of subtile, sublime, or delectable speculation, but such as shall be operative to the endowment and benefit of man's life. For it will not only minister and suggest for the present many ingenious practices in all trades, by a

74 Long, *Artisan/Practitioners*, 95.

75 Long, *Artisan/Practitioners*, 95. There is the caveat, as Long observes, that the phenomena of trading zones "did not occur universally through all society," nor did they change the "social/political hierarchy" (Long, *Artisan/Practitioners*, 128).

76 Paolo Rossi, *Philosophy, Technology, and the Arts in the Early Modern Era* (Torchbook, 1970), 58.

77 Elspeth Whitney, *Medieval Science and Technology* (Greenwood Press, 2004), 134.

78 Schatzberg, *Technology*, 42.

79 Eric Schatzberg has noted that Bacon's rejection of the Aristotelian divide between *technê* and *episteme* was accompanied by a desire for "the history of the mechanical arts to be an integral part of natural history," and to "subordinate the mechanical arts to scientific principles" (Schatzberg, *Technology*, 49, 51).

> connexion and transferring of the observations of one art to the use of another, when the experiences of several mysteries shall fall under the consideration of one man's mind; but further, it will give a more true and real illumination concerning causes and axioms than is hitherto attained. For like as a man's disposition is never well known, til he be crossed, nor Proteus ever changed shapes till he was straitened and held fast, so the passages and variations of nature cannot appear so fully in the liberty of nature as in the trials and vexations of art.[80]

Bacon's acceptance of the mechanical arts as beneficial for artisans practicing trades is much more than an acknowledgement of the value of manual labor. He underscores the intellectual value of the mechanical arts when they are used in the protean practice of translation, whereby the techniques applicable to one of the arts are applied in new and different contexts, a process of trial and error which comprises a feedback loop of improvement in the individual arts. A mind which would be able to comprehend the interconnectedness of various arts, based on techniques whose value transcends any particular one is also more likely to be able to understand the more general "causes and axioms" that would inform a theory of technical practice: what one might conceive of as part of the foundation of a technology.

A history of the arts also has a theoretical value for Bacon,[81] who continues to develop and expand his thinking about the mechanical arts in the *New Organon* (1620). Already the introductory aphorisms make a claim for the value of tool use, when Bacon states that "neither the bare hand nor the unaided intellect has much power" and that "the work is done by tools and assistance."[82] Other passages in the *New Organon* pick up on this theme, such as paragraph 85, which extols the virtue of the mechanical arts. Among other things, Bacon lauds the "the variety of objects," "the splendid equipment which

80 Francis Bacon. *The Advancement of Learning*, ed. William Aldis Wright. Fifth edition (Clarendon Press, 1920), 89–90.

81 Paolo Rossi describes how "The project of a history of mechanics, or history of the arts, was formulated for the first time in the *Advancement of Learning* (1605) and then taken up again on a much larger scale, and more forcefully, in the *Parasceve ad historiam naturalem et experimentalem* (published in 1620 as an appendix to the *novum organum*) and finally in *De dignitate et augmentis scientiarium* in 1623. The history of the arts, affirms Bacon, up to now has been considered in a way so wretched and useless that it must be placed among the *desiderata* of the new encyclopedia of the sciences" (Rossi, *Philosophy, Technology, and the Arts*, 118).

82 Francis Bacon, *The New Organon*, ed. Lisa Jardine and Michael Silverthorne (Cambridge University Press, 2000), 33.

the mechanical arts have assembled and contributed to human civilization," and the fact that at the heart of each is "subtle, ordered movement of hand and tool."[83] Within his praise for the mechanical arts, however, Bacon also embeds a critique. Things like the preparation of wine and beer, and the making of bread, are only in Bacon's day reaching a degree of perfection – even though they predate philosophy and the intellectual arts. If that is the case then, given that "these discoveries which we have just mentioned are older than philosophy and the arts of understanding," the truth of the matter is that "since such rational and dogmatic sciences came into being, the discovery of useful works has ceased."[84] To be fair, Bacon adds that the mechanic himself is partially to blame for he, "who is by no means anxious about the investigation of truth, does not direct his mind or stretch his hand to anything but what is useful for his task."[85]

4 Language Trouble

In addition to forcing a re-evaluation of the traditional lowly status suffered by the mechanical arts, already palpable in Greek culture and recodified in the classificatory schemes of knowledge established during the Middle Ages, Bacon points out an additional problem, one whose consequences reach well into Enlightenment-era discussions of technology: the problem of technical language. It is a well-known feature of Bacon's *New Organon* that he describes four classes of "idols" or "illusions": those of the tribe, the cave, the marketplace, and the theater. The third of these is of concern here. Bacon defines the idols of the marketplace as illusions which "seem to arise by agreement and from men's association with each other."[86] The reason is that "men associate through talk; and words are chosen to suit the understanding of the common people. And thus a poor and unskillful code of words incredibly obstructs the understanding."[87]

This initial definition, located in paragraph forty-three of the *New Organon*, does not refer to technical language per se. Instead, its parameters are defined in terms of class distinction – those who are involved with trade, who speak the "vulgar" languages of the masses. In the following paragraphs, however,

83 Bacon, *New Organon*, 69.
84 Bacon, *New Organon*, 70.
85 Bacon, *New Organon*, 81.
86 Bacon, *New Organon*, 41–2.
87 Bacon, *New Organon*, 42.

Bacon returns to the problem of language, connecting it directly to the sciences and, eventually, to the mechanical arts. The idols of the marketplace are "the biggest nuisance of all" because they impede our comprehension in various ways, due to the imposition of words on the understanding: "They are either names of things that do not exist (…) or they are the names of things which exist but are confused and badly defined, being abstracted from things rashly and unevenly."[88] Bacon refers here to concepts such as "fortune," "the first mover," "fire," and other "fictions."[89]

For Bacon, at least, the mechanical arts do better than the sciences where the use of language is concerned. Founded on nature and experience, these arts "constantly quicken and grow as if filled with spirit; at first crude, then adequate, later refined, and always progressing."[90] Not all early modern philosophers were in agreement about this, however. The seventeenth century demonstrates a certain degree of skepticism toward the language of the mechanical arts. One finds this skepticism clearly expressed by Gottfried Leibniz in a short essay titled "Unpresumptuous Thoughts Having to Do With the Use and Improvement of the German Language" (*c.*1697). Leibniz's essay is stamped by a historical perspective Bacon did not have access to – the aftermath of the Thirty Years War, which began in 1618 and ended in 1648, some twenty years after Bacon's death. The main point of Leibniz's essay is to emphasize that one of the greatest strengths of the German language, namely things pertaining to the arts and handcrafts, is also a source of rapid change. Such change is due no small part to the war:

> As the thirty years' war spread and gained the upper hand, Germany was overflooded with foreign and domestic peoples, as if by a deluge, and our language was not confiscated any less than our goods.[91] (my translation)

Leibniz's essay, written around 1697 and conceived of as a "handbook for German youth" [*Handbuch für Deutsche Jünglinge*] is somewhat of an anomaly in the oeuvre of this particular writer, both because of the subject matter and because of the fact that it was written in German, rather than Latin or French.

88 Bacon, *New Organon*, 48–9.

89 Bacon, *New Organon*, 49.

90 Bacon, *New Organon*, 61.

91 Gottfried Leibniz, "Unvorgreifliche Gedanken betreffend die Ausübung und Verbesserung der Deutschen Sprache" [Unpresumptuous thoughts having to do with the use and improvement of the German language] (Christian Georg Ackermann, 1831), 14.

It is also more than what the title itself suggests, because in addition to containing pragmatic ideas about the use and avoidance of technical terms, it offers several observations about the nature of language and what we do with it. The connection between these two aspects of Leibniz's essay – which can also be understood as the relationship between the economical use of language and the myriad empirical phenomena it necessarily encompasses – helps set the stage for the discussion that takes place about 100 years later.

Among its other contributions, Leibniz's essay provides its reader with a theory of words as signs that can stand for things as well as for thoughts. We manipulate these signs more "economically" according to Leibniz while speaking and thinking when we do not slow ourselves down with corresponding images for each and every word. Just as in the marketplaces one can pay with a piece of paper – a *Zeddel* or *Marcken* – rather than counting out individual coins, in order to make transactions move more quickly, one need not encumber oneself with the "images of things" [*Bildnessen der Dinge*] when the words are there to take their place. Stated somewhat more laconically: words are "the exchange-papers of the understanding" [*die Wechsel-Zeddel des Verstandes*].[92] One uses these signs to think and make progress in thinking, to find things out that the ancients did not know. At least, one tries to: language itself may also be "the mirror of the understanding" as Leibniz states in the first sentence of his essay, but German (at least, in his opinion) is not the most metaphysical language of them all.[93] It excels in "everything that can be grasped with the five senses" as well as "matters relating to the (technical) arts and handcraft," with particular strength in mining, hunting, forestry, seafaring, and "such things."[94] It is less good, according to Leibniz, at those things that are only reached at through contemplation or *Betrachtung*, including emotions (*Gemüths-Bewegungen*), theories of morality, logic, and metaphysics (i.e., those things with which the "common man" does not concern himself and which the educated person expresses in French or Latin).[95] What one has then, based on Leibniz's description, is a language that possesses a number of practical words and a corresponding deficit of the kind of abstract words necessary to generate the structural features of theory-building. That the acuity of this situation particularly affects technical terminology is evident in Leibniz's claim

92 Leibniz, "Unvorgreiffliche Gedanken," 47.
93 Leibniz, "Unvorgreiffliche Gedanken," 47.
94 Leibniz, "Unvorgreiffliche Gedanken," 47.
95 Leibniz, "Unvorgreiffliche Gedanken," 48.

that those words relating to governance (*Regierungs-Sachen*) and various civic matters pertaining to daily life and concerns of the state (*allerhand bürgerliche Lebens- und Staats-Geschäffte*) belong to the most significant deficits of the German language.[96]

Leibniz's response to the perceived inadequacy of German anticipates a phenomenon that will come to define eighteenth-century writing on technology and technical terms: in each case, one needs to establish genres of containment. But what form, precisely, should these genres take? A dictionary? Leibniz identifies no less than three possible kinds: one for common words, another for technical terms, and a third for older and provincial words. The genres themselves are established via translations from Latin and Greek: Leibniz refers to the first as a *Sprachbrauch* or *Lexicon*; the second as a *Sprach-Schatz* or *cornu copiae*, and the third as a *Sprachquell* or *Glossarium*. He sees models for such endeavors in the dictionary project of the Académie Française as well as Antoine Furetière's competing *Dictionnaire universel*, and in the Italian dictionary project undertaken by the group *La Crusca*. Leibniz was of the belief that a dictionary of technical terms, in particular a polyglottal one, would help generate "beautiful thoughts and inventions."[97] In other words: the dictionary of technical terms has a potential theoretical value.

Though unaware of the headaches his intellectual successors will face at the end of the eighteenth century, with the problem of containment Leibniz conjures a specter that will continue to haunt generations yet to come. That which for latter-day writers is a question of collecting technological words and objects, or defining technology as a science per se, is a problem Leibniz faces with regard to the German language, whose borders are remarkably – indeed, somewhat suspiciously – porous. On the one hand, this is a pragmatic question of which words have become legitimate "citizens" (as he puts it) in the chaotic linguistic aftermath of the Thirty Year's War. On the other hand, one can observe a tendency in Leibniz's essay to establish far-reaching territorial boundaries. After all, he argues, one can find German and Germanic words in many languages, including Greek, Latin, Persian, Swedish, Norwegian, and Icelandic. The joint questions – where does German end and where do other languages begin? – remain unanswerable in this context, but they will return, reconfigured, as a problem of technical terminology.

96 Leibniz, "Unvorgreiffliche Gedanken," 50.

97 Leibniz, "Unvorgreiffliche Gedanken," 59.

5 Technical Terminology in the Eighteenth Century

Significant aspects of eighteenth century life facilitated an interest in developing literary genres that could help order technical terminology. These include, for Daniel Roche, the "encounter between Cartesian modes of analysis and English empiricist perspectives," "the rise of utilitarianism in the light of which inventories of knowledge could be perceived as indispensable to the progress of society," and "new standards of cultural sociability in Europe" that facilitated a "perceived conjunction of institutions of knowledge and power."[98] At the basis of technical progress, knowledge, and societal institutions was language, and technical language was increasingly identified as a problem in the eighteenth century. What some have referred to as "Europe's first alphabetical encyclopedia of business,"[99] the *Dictionnaire universel de commerce* of 1723–1730, tried to sort through "technical keywords" and "neologisms" that "referred, among other things, to commercial conventions and documents, specialized workers, parts of machines, tools, and processes" as well as "regulations and institutions."[100]

Diderot and D'Alembert encountered this problem in the context of their massive encyclopedia project, published between 1751 and 1772, which tried to systematically order information about the current state of the sciences, trades, and arts (broadly understood). They subtitled their project *dictionnaire raisonné*, or "analytical dictionary," in which capacity it should, in Daniel Brewer's words, provide "definitions that stated the general principles and essential details pertaining to all branches of science and the liberal and mechanical arts of the eighteenth century."[101] As an encyclopedia, the ambitions of the

98 Daniel Roche, "Encyclopedias and the Diffusion of Knowledge," in *The Cambridge History of Eighteenth-century Political Thought*, ed. Mark Goldie and Robert Wokler (Cambridge University Press, 2006), 173.

99 André Morellet, paraphrased by Jeff Loveland in Loveland in "The *Dictionnaire universel de commerce* (1723–1730)," in *Specialized Dictionaries and Encyclopedias, 1650–1800*, ed. Jeff Loveland and Stéphane Schmitt (Liverpool, 2024), 289.

100 Loveland, "The *Dictionnaire universel de commerce*," 301.

101 Daniel Brewer, "The *Encylopédie*: Innovation and Legacy," in *New Essays on Diderot*, ed. James Fowler (Cambridge University Press, 2011), 47–8. With regard to the status of the mechanical arts in the eighteenth century, Bert De Munck has commented on a transformation in their value: even if "technical knowledge and hands-on skills were seen as useful," he writes, "artisanal skills were henceforth detached from knowledge and truth" as they "no longer gave access to God's wisdom and hence the secrets of the universe." Bert de Munck "Shared Skills and Technologies of Community Formation. Artisanal Epistemologies, Secrecy, and Governmentality in Long-Term Context," in *The Republic of Skill. Artisan Mobility, Innovation, and the Circulation of Knowledge in Premodern Europe*,

project were to "establish and display the fundamental interconnectedness of all forms of human knowledge."[102] "All forms" included the mechanical arts, which required "plumbing the depths of a commercial world deemed unfit for serious scrutiny in the age of classicism."[103] Diderot and D'Alembert's research led them to the older tradition of "practical treatises on arts and crafts" inspired by the "sixteenth- and seventeenth-century *théâtre des machines*."[104]

In the course of their research, Diderot and D'Alembert encountered difficulty on the micro level: that of words. Diderot's entry on "art" puts the problem in no uncertain terms: "I have found the language of the arts to be very imperfect for two reasons," he writes, "the scarcity of proper nomenclature and the frequency of synonyms."[105] A few lines down, he adds, "One would wish for more attention to analogy of form and use."[106] D'Alembert, for his part, takes a somewhat different angle in the "Preliminary Discourse" to their encyclopedia project. There he writes that the sheer quantity of technical terms is so vast that a "man of letters" would likely not know one twentieth of the existing words, because "although each art has its own language, this language is still very imperfect" and that "it is through the long-established habit of conversing with one another that the workers make themselves understood."[107] In fact, D'Alembert believed that when craftsmen make themselves understood, "this is accomplished much more by the repetition of contingent actions than by the use of terms" since "in a workshop it is the moment that speaks, and not the artisan."[108] The question of to what degree workshop practices involve knowledge-making has also been considered as a problem of material culture for scholars of the Renaissance and seventeenth centuries. Celina Fox

ed. David Garrioch (Brill, 2022), 47. In support of this view, De Munck also notes that in Diderot and D'Alembert's encyclopedia project, "artisans were rendered as automatons, sophisticated robots who were able to make but not invent beautiful and valuable things."

102 Brewe, "The *Encylopédie*: Innovation and Legacy," 48.

103 Roche, "Encyclopedias and the Diffusion of Knowledge," 27.

104 Roche, "Encyclopedias and the Diffusion of Knowledge," 27.

105 Denis Diderot (attributed), "Art," in *The Encyclopedia of Diderot & d'Alembert Collaborative Translation Project*, trans. Nelly S. Hoyt and Thomas Cassirer (University of Michigan Library, 2003), http://hdl.handle.net/2027/spo.did2222.0000.139 (accessed August 16, 2023). Originally published as "Art," *Encyclopédie ou Dictionnaire raisonné des sciences, des arts et des métiers*, 1:713–717 (Paris, 1751).

106 Diderot (attributed), "Art" (accessed August 26, 2023).

107 d'Alembert, Jean-Baptiste le Rond, "Preliminary Discourse," in *The Encyclopedia of Diderot & d'Alembert Collaborative Translation Project*, trans. Richard N. Schwab and Walter E. Rex. (University of Michigan Library, 2009), http://hdl.handle.net/2027/spo.did2222.0001.083 (accessed August 26, 2023). Originally published as "Discours Préliminaire," *Encyclopédie ou Dictionnaire raisonné des sciences, des arts et des métiers*, 1:i–xlv (Paris, 1751).

108 D'Alembert, "Preliminary Discourse" (accessed August 26, 2023).

has discussed how these investigations reveal "rich repositories of empirical knowledge – the knowledge of nature and matter arising from artisans' skills" to the degree that "workshop practices have been exposed not as the mindless following of rules or recipes but as processes of knowledge-making that involved extensive experimentation and observation."[109]

In the eighteenth-century French context, at least, Diderot and D'Alembert were not alone in their diagnosis that there was a problem with technical language. The first volume of a new edition of the *Descriptions of Arts and Trades*, published in 1771 through the initiative of the Parisian Royal Academy of Sciences, under the guidance of Jean Élie Bertrand, contains a preface written by Bertrand that goes into somewhat more detail:

> One of the great difficulties in the description of the arts is the very language of the artisans. Each art has a multitude of technical terms and manners of speaking analogous to the operations, the tools, and the machines being used. These terms are absolutely necessary for making oneself understood in the ateliers and describing the handiwork (*manoeuvres*). But philosophers are not the ones who have invented this language. In each art one has multiplied the terms; from one to the next, one has changed them too much. Each [art] has its particular vocabulary, too often very different from one to the next, for things which, however, are common to all, or at least to several. To vary the terms without necessity, does it not place an obstacle to the progress of mechanical knowledge?[110]

Other sciences do not escape Bertrand's critique. The natural sciences, for example, also display the same kind of "tiring polynomy" that the mechanical arts do. It is the latter, however, to which Bertrand directs both his critical energy and concrete suggestions for reform. Physical appearance, the circumstances of use, and the "identity of maneuvers" – how one uses them – should be enough of a reason for different tools to have the same designation.[111] The end effect, then, would be a "more uniform nomenclature."[112]

The question of technical terminology resonated in the eighteenth-century German context as well, and fell squarely under the auspices of what, in the

109 Fox, *The Arts of Industry*, 5.

110 Jean Élie Bertrand, "Préface," in *Descriptions des arts et métiers* [Descriptions of arts and trades], New edition, vol. 1 (Société Typographique, 1771), xi.

111 Bertrand, "Préface," xii.

112 Bertrand, "Préface," xii.

1770s, received the designation *Technologie*. Johann Beckmann, in the introduction to his *Guide to Technology* from 1777, comments that "it is unpleasant that some tools and works, in various handcrafts, have entirely different designations," adding that "if one wanted to work philosophically or systematically on technical terminology, one would be required to do away with more synonyms than to introduce new names."[113]

6 From Technical Terms to Technology, the Problem Word

Just as technical terms present difficulty, "technology" itself has historically been perceived as a troublesome word. If one had to categorize the kind of trouble that technology engenders, at least three areas of irritation come to mind: 1) disagreement about when technology acquired its modern meaning (with the concomitant question of what this modern meaning is); 2) disagreement about whether it is correct to refer to "technologies" of any sort in historical settings before the concept of technology existed; 3) disagreement about how *Technologie* functions in the German context from the 1770s through the mid-nineteenth century. Since the latter of these will be addressed more extensively in the essays accompanying the translations, just a brief overview of the first two areas of disagreement will be provided in order to give a sense of how far-reaching the problem actually is.

With regard to the question of when technology acquired a "modern" meaning – leaving aside for a moment the disagreements about what that modern meaning actually is – it is possible to find answers ranging from the seventeenth through the twentieth century. As always, the devil is in the details. Ortega y Gasset, for example, links the advent of "modern technology" to the development of a mechanical philosophy: "technology properly speaking, technology in the fullness of its maturity, begins around 1600, when man in the course of his theoretical thinking about the world comes to regard it as a machine."[114] It begins, according to this perspective, with Galileo, Descartes, and Huygens and their "mechanical interpretations of the universe."[115] Others, using different metrics, see the modern meaning of technology arriving at later historical moments. When James M. van der Laan writes that "the word 'technology' acquired its more or less modern meaning only a few centuries ago,"

113 See p. 73.

114 José Ortega y Gassett, *Toward a Philosophy of History* (University of Illinois Press, 2002), 117.

115 Ortega y Gasset, *Toward a Philosophy of History*, 117.

he chooses John Harris's *Lexicon Technicum* from 1704 as his point of reference, given that it "must be considered the first technical dictionary."[116] Harris's definition of the technical is as follows:

> Technical is sometimes the same with *Artificial*, and expresses whatever relates to the Arts and Sciences, as the *Terms, Rules,* &c. So that the Terms of Art are commonly called *Technical Words*.[117]

Van der Laan's own commitment to this reference point contains a degree of skepticism – he offers the subsequent caveat that Harris "anticipates" but perhaps "does not yet comprehend" the modern meaning of technology.[118] From the point of view of historical sources, the softer position is certainly safer: there are a number of English-language dictionaries predating Harris's that offer essentially the same equation of the technical with the artificial, such as Edward Phillips' *New World of English Words* from 1658, Elisha Coles' *English Dictionary* from 1676, and Thomas Blount's *Glossographia* from 1681. Several of these dictionaries, such as Coles', also contain an entry on "technology," where it is defined in the older sense of "a treating of arts or workmanship."[119]

Those who see a later date for the arrival of a modern understanding of technology often look towards the time period of increasing industrialization in conjunction with advances in the sciences. Here, too, the particular historical vantagepoint matters. Vincent Leitch might write, from the perspective of a "postmodern theory of technology," that "by common consent *modern* technology starts during the Enlightenment period when the widespread scientization of technology and technologization of science occur," but the implication is that modern technology essentially begins … when modern technology begins.[120]

The recent work of Leo Marx, Eric Schatzberg, and others has both located the emergence of a modern understanding of technology toward the end of the nineteenth century and, more importantly, given a theoretical justification as to why that is the case. Leo Marx, in his essay on technology as a "hazardous concept," acknowledges the arrival of technology in seventeenth-century English language, but looks ahead, historically, to identify the "specific changes

116 J. M. Van der Laan, *Narratives of Technology* (Palgrave Macmillan, 2016), 4.

117 John Harris, *Lexicon technicum: or, an Universal English Dictionary of Arts and Sciences*, vol. 1. (Dan Brown, 1704), unpaginated.

118 van der Laan, *Narratives of Technology*, 4.

119 *An English Dictionary*, Elisha Coles, 1st edition (Samuel Crouch, 1676), s.v. "Technology."

120 Leitch, "Postmodern Theory of Technology," 209.

that prompted the emergence of *technology* – the concept, the word, the purported thing itself."[121] According to Leo Marx, "those changes … created a semantic – indeed, a conceptual void, which is to say, an awareness of certain novel developments in society and culture for which no adequate name had yet become available."[122] In other words, a perceived discursive lack precipitated the invention of the concept. The "hazardous" aspect Leo Marx describes stems from the transformation of the mechanical arts in to "an ostensibly discrete entity – one capable of becoming a virtually autonomous, all-encompassing agent of change."[123] The linguistic aftershocks of such a transformation include our ability to speak of things such as technology's "acceleration," as if technology had its own independent agency – something that clearly was not a feature of the discourse on technology around 1800. Certainly there are those who would point, as an example, to passages in the work of Karl Marx. A well-known footnote in his discussion of the production of the relative surplus value (*Mehrwert*) describes how technology "reveals man's dealings with nature, discloses the direct productive activities of his life, thus throwing light on social relations and the resultant mental conceptions."[124] Karl Marx's understanding of "technology" here is clearly a marked departure from early Enlightenment definitions equating technology with a collection of rules governing the practice of the arts. For Leo Marx's purposes, however, the existence of such isolated eruptions of a more modern take on technology do not amount to much because they aren't gathered systematically into Karl Marx's larger philosophical project.

One aspect of Leo Marx's work to establish the emergence of a modern definition of technology at a particular historical moment involves a critique of the idea of a "history of technology," the fact that even though the word technology is very late to arrive, "the belated emergence of the word [is] used to name the very rubric."[125] Cuomo's *Technology and Culture in Greek and Roman Antiquity*, van der Laan's *Narratives of Technology*, Long's *Technology, Society, and Culture in Late Medieval and Renaissance Europe*, Ovitt's work on *Labor and Technology in Medieval Culture*, and Whitney's *Medieval Science and Technology* are just a few of the many historical studies of technology whose invaluable intellectual contributions to our understanding of earlier cultural practices are required,

121 Leo Marx, "Technology. The Emergence of a Hazardous Concept," in *Technology and Culture* 51 (July 2020): 563.

122 L. Marx, "Technology," 563.

123 L. Marx, "Technology," 563.

124 Karl Marx, *Capital* (Dent, 1930), 393.

125 L. Marx, "Technology," 562.

out of sheer necessity, to be articulated under auspices of a concept that, in the pertinent time period, does not yet exist. In keeping with the spirit of the epigraph to this introduction, however, we don't need to make this an intellectual misdemeanor, but it is something to keep in mind when speaking about technology in historical contexts. The present project manages to avoid such a critique by taking a close look at how the word *Technologie* assumes many roles around 1800. One of the central arguments here is that *Technologie* was more than a simple collection of rules designed for application in the arts. The question is: what constitutes this "more"? The following overview will give readers a general sense of what is at stake, and the essays that accompany the translations contained in this volume will go into more detail about the specific nuances of *Technology* as understood in the translated texts.

As a German word, *Technologie* is, in the eighteenth century, most frequently associated with cameralism, an economic term that emerged in seventeenth-century German principalities, and whose primary concerns were "directed to the state's interest in its resources, in better administration and in the common good, the purpose being in order to increase the prince's incomes, establish a sustainable development of economy, and create a well-ordered state."[126] Beckmann writes in the preface to the first edition of his *Guide to Technology* that "the knowledge of handcrafts, fabricks, and manufactured goods is indispensable to everyone who wishes to dedicate himself to statecraft and the science of cameralism."[127] The administrators-in-training of the German principalities were supposed to have a good working knowledge of both the liberal and the illiberal arts. Yet cameralism was only the latest resting

126 Marten Seppel, "Cameralism in Practice," in *Cameralism in Practice. State Administration and Economy in Early Modern Europe*, ed. Marten Seppel and Keith Tribe (England: Boydell and Brewer, 2017), 1. Seppel notes that cameralism was "first and foremost a way of thinking and a common language (*rhetori*)," and that there is "little consensus over the specific and definite character of 'cameralism'" (Seppel, "Cameralism in Practice," 5–6). Keith Tribe also underscores the notion of cameralism as a doctrine of teaching, recorded in books for academic use when he writes that "cameralism was a form of academic pedagogy" and that "(a)s a written discourse it was embodied in the several hundred textbooks produced for use by students in German, Austrian, and Baltic universities between the 1720s and the 1790s." See Keith Tribe, "Cameralism and the Sciences of the State," in *Cambridge History of Eighteenth-Century Political Thought*, ed. Mark Goldie and Robert Wokler (Cambridge: Cambridge UP, 2006), 525. Andre Wakefield has critiqued Tribe's interpretation of cameralism with the concern that historians have tried to fit this "extinct" science into their own frameworks of knowledge without understanding cameralists "on their own terms." See Andre Wakefield, *The Disordered Police State. German Cameralism as Science and Practice* (The University of Chicago Press, 2009), 5.

127 Johann Beckmann, preface to the first edition of the *Guide to Technology*, p. 58.

point of a long discursive migration of the word "technology" since antiquity. For the Greeks, the cognate word referred to a systematic treatment of something, such as a grammar,[128] which is also the sense in which Cicero borrowed it for Latin: one can think of his *de Oratore* as a technology of rhetorical technique. The defining element of technology as it is used in this earlier sense is a collection of rules designed for application. One still finds the understanding of technology as a collection of rules through the seventeenth century, where its application was expanded to include other areas of study, such as theology; Gaspard Streso's 1633 *Technologia Theologica*, for example, applied rules of logical analysis to theological tenets. As far as the German context is concerned, Christian Wolff provides a good point of reference for understanding the early-eighteenth century discussion on technology at the cusp of Latin and German as the scholarly language of choice. In his *Philosophia rationalis sive logica* (1728), Wolff offers this simple, yet programmatic definition:

> It is also possible to have a philosophy of arts, even if this has been hitherto neglected. You can call it technics (*technical*) or technology (*technologia*). Therefore, technology is the knowledge (*sciencia*) of arts and of works of art, or, if you prefer, the knowledge of those works which are completed by men with the instruments of their body, especially the hands.[129]

Seibicke has argued that "with this special use of both expressions (*technica* and *technologia*)," Wolff laid "the basis for the development of the concept of technology (*des Technikbegriffs*) so important for modernity," and that one can see this as the first step in the "program of a systematic science of technology (*Technikwissenschaft*)."[130] Certainly, several aspects of Wolff's definition bear directly on later eighteenth-century usage: first, technology deals both with the knowledge (or science) of the arts and with the products of that knowledge – the works of art themselves. "Art," in this context, is understood

128 In a study devoted to the "word family" *techne* in the German language, one of Wilfried Seibicke's many contributions is to problematize the priority of technology's two key semantic components by pointing out that various historical time periods have witnessed the predominance of a "logos der techne" as opposed to a "techne des logos." See Wilfried Seibicke, *Technik. Versuch einer Geschichte der Wortfamilie* techne [Attempt at a history of the word family techne] (VDI-Verlag, 1968), 97.

129 Christian Wolf, *philosophia rationalis sive logica* (Rational philosophy or logic), 2nd edition (Renger, 1732), 33.

130 Wilfried Seibicke, "Von Christian Wolff zu Johann Beckmann," in *Mehrsprachigkeit in der deutschen Aufklärung*, ed. Dieter Kempel (Meiner, 1985), 43.

in the broad sense of man-made things before being specified to works of the hand, but otherwise makes no distinction between the fine and banausic arts. Secondly, this definition articulates an important point of bifurcation. Wolff offers a choice between *technica* and *technologia* – as if to say that we could conceivably have a philosophy of arts without the "logos" which comprises scientific discourse – before leaning towards the latter. Over time, the distinction between technics and technology takes different forms, including the immediacy of social practices versus the ideas which encompass them.[131]

Some idea of technology is therefore already active in the German context when Johann Beckmann writes, in the fourth edition to his *Guide to Technology* (1796), "I first dared in 1772 to use technology (*Technologie*) instead of the customary designation history of arts (*Kunstgeschichte*), which is at least as incorrect as the designation history of nature (*Naturgeschichte*) for the theory of nature (*Naturkunde*)."[132] A few of Beckmann's contemporaries, such as Daniel Heinrich Bensen, professor of Cameral Sciences in Erlangen in the 1790s, found Beckmann's statement problematic: "Why does one not prefer, if one is in agreement on the general concept of a handcraft (*Handwerk*), to use the German word *Handwerkskunde* (study of handcrafts) which simply and precisely expresses the exact same thing?"[133] With the advantage of hindsight, later historians such as Seibicke will claim that Beckmann's *Guide to Technology* "marks the change from the quantity of techn(olog)ical knowledge into the new quality of a techn(olog)ical science."[134] Whether or not one is willing to go that far, Beckmann does go at least a step farther than Wolff's notion of a "science of arts and works of art" when he articulates the distinction between *Technologie* and *Kunstgeschichte*. According to Beckmann, a history of art (that is, of the arts in general) might describe the invention and fate of a particular craft, but "technology is something which explains, in an orderly manner, and which establishes the foundation and logical consequence of all kinds of works."[135] The tension inherent in the eighteenth-century discussion on technology does not just emerge as a distinction between Wolff's and Beckmann's approaches, however. There is also a third source to bring into this conversation, one which

131 Mitcham, *Thinking through Technology*, 276.

132 Johann Beckmann, *Anleitung zur Technologie*, 4th edition (Vandenhoeck and Ruprecht, 1796), 20.

133 D. Heinrich Bensen, *Versuch eines systematischen Grundrisses der reinen und angewandten Staatslehre für Kameralisten* (Attempt at a systematic outline of pure and applied political theory) (Erlangen: Johann Jacob Palm, 1798), 204.

134 Seibicke, "Von Christian Wolff zu Johann Beckmann," 47.

135 Ibid.

stands for another problem inherent in the emerging reflections on technology: Johann Heinrich Zedler's *Universal Lexicon*.

The significance in Beckmann's act of translation has less to do with a linguistic transferal among languages than a translation from one way of identifying a field of thought to another. By contrast, the definition of *Technologie* in Zedler's lexicon – a massive, sixty-eight volume project published between 1731 and 1754 – is based on words. For Zedler, technology is synonymous with *Kunst-Wörter-Lehre*. As a doctrine of *Kunstwörter* (where *Kunstwort* indicates a technical term), technology becomes a way of organizing things that belong to people of a particular trade. Along these lines, a carpenter will have specific words for his tools and techniques, as will all other craftsmen. To the degree that Zedler's definition of "technology" suggests a collection of these words, it is situated in the classical[136] rhetorical tradition of Cicero. Yet even if Zedler's definition is clear enough, the question of what it denotes poses a problem for the German language. The technical terms of the different arts often required mastery of Latin, which led to the common practice of translating the technical terms into German somewhat arbitrarily, "without the approval of others" so as to "cause obscurity."[137] In other words, based on Zedler's lexicon, it is clear that the early translation of technology as a concept from Latin to German carries with it the problem of translation in general – at least as a problem of the parts (i.e., the technical terms), if not yet the whole.

Against the backdrop of Wolff and Zedler, as well as other eighteenth-century sources, Beckmann's own "translation" of technology therefore takes a different shape altogether, beginning with a re-definition of his point of departure. Because Beckmann understands himself as departing from a history of the arts, he requires a theoretical apparatus, something closer to a science. This idea is fundamental to the definition of technology found in the first edition to his *Guide to Technology* (1777) where he writes that "technology is the science which teaches the refinement of natural products ... the knowledge of handcrafts."[138] What Beckmann has in mind is inspired as much by a "philosophy of the arts" as it is by Zedler's understanding of technology in terms of a doctrine of technical terms, albeit with a few key differences. Having

136 Wolfhard Weber shows how Beckmann's choice of the word *Technologie* also might have been influenced by French sources, since an early usage of the adjective "technological" appears in the context of a discussion of French wallpaper manufacturing (Weber 1980, 146).

137 *Universal-Lexicon*, Johann Heinrich Zedler, 1st edition, vol. 42 (Leipzig: Johann Heinrich Zedler, 1744), s.v. "Technologie."

138 Beckmann, preface to *Guide to Technology*, 1st edition (1777), p. 75.

established in not-too-specific language that technology is a science that handles potentially everything in nature,[139] a second part of the definition goes a step further to define the place of the discourse of technology in society in contradistinction to the physical place of the workshop:

> Instead of just being shown in the workshops how one should follow the guidelines and customs of the master for the fabrication of wares, technology gives, in systematic order, thorough instruction as to how one finds the means from true principles and reliable experiences for precisely this goal, and how one should explain and use them for the refinement of the phenomena at hand.[140]

If technology, according to Beckmann, can act in the place of the master craftsman who gives instructions in the workshop in that it provides a "systematic order" and "thorough directions" based upon both principles and practice, then it is more than an empirical science. Beckmann's technological project implies a collapse of both discursive and spatial distinctions: that is, the collapse of a difference between theory and practice, as well as the difference between the work (and place) of the scholar and that of the craftsman. And although Beckmann wants to establish his "technology" in a trajectory apart from Zedler's by claiming it as a science whose system achieves validity to the degree that it is not taking part in a history, he must nonetheless confront the technology of words as posed by Zedler by re-scripting the space of the workshop, with its profusion of technical terms, its clash of vernaculars. Through Beckmann's imagined projection into the space of the workshop, two technologies – the one, a science of arts and the other, a theory of words – become deeply intertwined.

To illustrate the connection between technological lexica and systems of technology more clearly, and to underscore the problems of both as pedagogical topics, one can take as an example Beckmann's preface to Jacobsson's "Technological Dictionary" (1781–84).[141] With regard to the former: Beckmann

---

139 More recently, Antoine Picon has encouraged historians of technology to remember that, even though it is often defined as "a form of social production," it "must not be forgotten" that technology "provides an interpretation of nature in relation to the social divisions of labour" and is "meant to adapt this division of labour to the principles of efficiency that can be deduced from the observation of the physical world," an idea that is very much in line with Beckmann's own approach (Picon 1996, 39).

140 Preface to Beckmann, *Guide to Technology*, 1st edition (1777), p. 75.

141 Georg Lamprecht, *Lehrbuch der Technologie oder Anleitung zur Kenntniß der Handwerke, Fabriken und Manufacturen* [Instruction book of technology or guide to knowledge of

writes that what one can expect to find in a technological dictionary such as Jacobsson's is not just the technical terms that refer to the methods of production unique to the various arts, but also those words which refer to materials used by artisans, different kinds of crafts, and the wares produced by them. Much of Beckmann's admiration for Jacobsson comes from the enormity of this task, whose difficulty is only increased by the number of synonyms and dialect words for any given term; the number of crafts alone is "so great, that it is already difficult just to compile a complete index of them" not to mention the tools and materials unique to each.[142] The comparison which Beckmann draws from throughout the preface to the dictionary, and which dates back to his earlier work, is with the study of nature. Already in the *Guide to Technology* he invoked the name of Carl Linnaeus, the eighteenth-century botanist renowned for generating a system and descriptive technical language to organize his far-ranging observations of natural phenomena. The proliferation of crafts and the words of human artifice is second only to the number of phenomena in the natural world. Yet the analogy is not only between two scientific discourses of comparable enormity. Beckmann's preface also insists that in order to understand the multitude of words drawn from the most various kinds of crafts, we need at our disposal a set of auxiliary sciences (*Hülfswissenschaften*) in addition to the theory of nature, including mineralogy, chemistry, metallurgy, and mathematics.[143] Which is to say that the "parts" of technology, in addition to being a collection of words, are also a collection of mutually implicated discourses and disciplines.

Technological writing around 1800 is deeply, unapologetically embedded in historical thinking. Beckmann's replacement of *Kunstgeschichte*, as a history of the arts, with the term *Technologie* is, in some ways, an emancipatory gesture, but it does not signal a move to de-historicize the idea of what technology is and can be around that time. For the purposes of the present study it does not make sense to consider historical narrative and speculative thinking as orthogonal to one another. The translations will show that what at first glance seem to be straightforward historical overviews, such as Johann Poppe's idea of a

---

the handcrafts, fabricks, and manufactures] (in der Hemmerdeschen Buchhandlung, 1787). Compared to Beckmann, relatively little has been written on Lamprecht, whose professional title at Halle was Professor der Rechte und Philosophie; he is referred to by a (near) contemporary as "the first who, through his draft of an encyclopedia and methodology of economic, political, and cameral sciences for use in academic lectures … sought to remedy a gap." Georg Gottfried Strelin, *Versuch einer Geschichte und Literatur der Staatswissenschaft* (Johann Jakob Palm und Ernst Enke, 1827), 49.

142 Beckmann, preface to *Guide to Technology*, 1st edition (1777), p. 105.

143 Beckmann, preface to *Guide to Technology*, 1st edition (1777), p. 105.

history of technology, encompass moments of reflection that transcend attention to particular historical moments.

The translations also testify to the emergence of two kinds of taxonomy of technology at the end of the eighteenth century. The first of these, based on a distinction between "chemical" and "mechanical" technology, is purely descriptive, an attempt to distinguish between different processes of making. The second one relies on a more abstract distinction made between "general" and "special" technology. Beckmann's 1806 "Draft of the General Technology" is instructive in this regard, but the distinction already arises in the introduction to Georg Lamprecht's *Instruction Book of Technology* (1787), in a chapter of reflections on the "concept of technology." For Lamprecht, at least, the defining characteristic of a "general technology" is its expansiveness, its ability to encompass "the theories of all kinds of handcrafts in a narrower sense that exist in the world, without distinction."[144] "Particular technology," in his conceptual framework, has a cultural/regional connotation. For example, European or German technology would be considered a particular technology. Beckmann does it differently, preferring to maintain the distinction between a general and particular technology as one that exists between the whole and the parts, whereby the latter are understood as "the description of individual handcrafts."[145]

7 The Organization of the Project

The first section, "From *technologia* to *Technologie*," charts the arrival of *Technologie* in the German language well before systems of technology were developed. It opens with a translation of the entry *Technologie* in Zedler's *Universal Lexicon* (1744) in order to provide a backdrop for the innovative thinking about technology found in the work of Johann Beckmann. The following two translations from the preface and introduction to the first edition of Beckmann's *Guide to Technology* (1777) illustrate how Beckmann defines his foundational terms – art, craft, and trade – and also reveal how he positions his work with regard to earlier projects such as the Diderot and Dalembert's encyclopedia. The translations of the second section, "Technological Orders," begin with the two prefaces to Johann Jacobsson's *Technological Dictionary* (1781) written by Beckmann and Jacobsson, before introducing a new contributor to technological thinking around 1800: Georg Friedrich von Lamprecht (1759–1810). The

144 Lamprecht, *Instruction Book* (1787); p. 120.

145 P. 166.

introduction to Lamprecht's *Instruction Book of Technology* offers an alternative approach to Beckmann's where the organization of technical objects and practices is concerned. The final section, "Universal vs. Historical Technology, circa 1800," makes evident the kind of tension between the theoretical framework governing historical narratives of technology and attempts to "universalize" what technology is around 1800. It contains a complete translation of Beckmann's "Draft of the General Technology" (1806) and Johann Poppe's "General Introduction to the History of Technology" (1807). As Genevieve Warwick and Richard Taws have noted, any historical study of technology "is not only concerned with practice-based forms of knowledge, but also with the changing conceptual paradigms that drive technological change" and any analysis of technology joins *homo faber* with *homo sapiens*, "the human manufacture of tools and instruments as the material extensions of productive bodily labour, but also patterns of thought and intention."[146] The following translations will allow readers to get a close-up look at how thinking about technology, the work of *homo sapiens*, emerged from a milieu of practical, financial, and intellectual concerns about the labor of *homo faber*.

8 *Technologie* after Beckmann, Lamprecht, and Poppe

This study is bookended historically by two questions which, each in its own way, are challenging to answer: what are the implications of Johann Beckmann's preference for *Technologie* in the 1770s, and what happened to German thinking about *Technologie* after the first decade or so of the nineteenth century? With regard to the latter, there are various paths one could pursue. For example, the data on the teaching of technology at German universities shows that student attendance in technological lectures in the context of the "cameralist sciences" peaked around 1810 before quickly declining.[147] Hans Frambach, Keith Tribe, and others have also described how cameralism came to be separated

146 Richard Taws and Genevieve Warwick, "After Prometheus: Art and Technology in Early Modern Europe," in *Art and Technology in Early Modern Europe*, ed. Richard Taws and Genevieve Warwick (Wiley-Blackwell, 2016), 16.

147 Weber, for example, writes that at the end of the eighteenth century there were approximately 32 professorships at German universities devoted to technology and related disciplines, and that around 45% of university students attended lectures on Cameralist sciences. See Wolfhard Weber, "Technik zwischen Wirtschaft und Handwerk. Die Technologie des 18. Jahrhunderts als Lenkungswissenschaft des spätabsolutischen Staates," in *Wirtschaft, Technik und Geschichte: Beiträge zur Erforschung der Kulturbeziehungen in Deutschland und Osteuropa*, ed. Albrecht Timm, Volker Schmidtchen, and Eckhard Jäger

from questions of the national economy: "whereas the former was reduced to the theory of finance and *Polizey*, the latter was concerned with questions of wealth."[148] As cameralism disappeared from university curricula, it became "a mere appendix to the study of law."[149] Robert Multhauf, who writes about the historiography of technology, also notes how historical works on *Technologie* dwindled during the nineteenth century and has inquired, why, as a history of technology of substantial scope, Poppe's work "remained almost unique for a century and a half, during which nearly everyone forgot that it existed," particularly since "this was a period during which technology flourished as never before."[150] It is also well-known that Karl Marx read and took notes on writings by Poppe and Beckmann, yet equally acknowledged that Marx only rarely refers to *Technologie* in his writings.[151]

A sample of books with *Technologie* in their title, published in the years following Beckmann's 1806 essay on general technology and Poppe's 1807 historical account, can also reveal a few clues as to why the discourse surrounding *Technologie* failed to flourish even as German industry ramped up, and why "technics" (*Technik*) took center stage.[152] Karl Christian von Langsdorf's *Erläuterung höchstwichtiger Lehren der Technologie* [Explanation of the most important doctrines of technology] from 1807 is a good point of departure, both because it was published the year after Beckmann's general technology, and because his fairly conservative standpoint (technology is "concerned only

(Camen 1980), 147. This number dropped to 7% by 1860, whereby interest in the natural and philological sciences increased (Weber, "Technik," 150).

148 Hans Frambach, "The Decline of Cameralism in Germany at the Turn of the Nineteenth Century," in *Cameralism in Practice. State Administration and Economy in Early Modern Europe*, ed. Marten Seppel and Keith Tribe (The Boydell Press, 2017), 257. Keith Tribe also comments how the the nineteenth century witnessed "cameralism's function displaced by faculties of law and its substance by *Nationalökonomie* [national economy]" (Tribe, "Cameralism and the Sciences of the State," 525). Andrew Jamison has also shown how Ure synthesizes two perspectives, the "innovation theory that focused on technology in the economy" and another "that focused on technology in its own right" (Jamison, "Technology's Theorists," 512).

149 Frambach, "Decline of Cameralism," 259.

150 Robert Multhauf, "The Historiography of Technology: Some Observations on the State of the History of Technology," *Technology and Culture* 15.1 (1974): 2.

151 For more context, readers can refer to Donald MacKenzie's essay, "Marx and the Machine" in *Technology and Culture*, vol. 25, no. 3 (Jul., 1984) 473–502.

152 To see these differences mapped out more dramatically, readers are encouraged to do a Google Ngram search using the terms "Technologie, Technik" between the years 1800 and 1900.

with material objects") provides a good foil for the more experimental thinking of those who come afterward.[153]

In the first decade or two after Beckmann's general technology, one can see various attempts to work with categorical distinctions already in place (such as, for example, the distinction between chemical and mechanical technology, or between theories of nature, such as one finds in a natural history, and technology). In this vein, one finds Poppe's self-proclaimed "first instruction manual of general technology" from 1809 in which he, like Beckmann, conceives of a general technology in terms of certain kinds of technical processes (splitting, reducing, joining, etc.) under the rubric of "giving form."[154] Poppe also advocated for a more economical use of tools and techniques among the handcrafts. Without overstating the case, one can see that Poppe (like Beckmann before him), is not "only" concerned with material objects, but also with more abstract questions: how to categorize types of techniques, for example, and evaluate them in terms of their relative usefulness and efficiency.

In the following years, one also finds attempts to think more generally about the concept of technology. Sigismund Hembstädt, for example, who takes as his point of departure the work of Beckmann and Poppe, creates a distinction between a "higher" and a "lower" technology, whereby the lower one "comprises the principles of general economy" and the higher one "comprises the embodiment or essence (*summarischen Inbegriff*) of all principles of the 'lower technology' along with its fundamental and auxiliary sciences."[155] One of the more unusual contributions, however, comes from Karl Karmarsch, who helped found the polytechnic school in Hannover, Germany, which later became the university of Hannover. Karmarsch's focus is on mechanical theory as it relates to technology. A preface written by his former adviser in Vienna, Georg Altmütter, does the polemical work of separating Karmarsch from Beckmann. It refers to the discussion of technology of twenty years past as nothing other

153 Karl Christian von Langsdorf, *Erläuterung höchstwichtiger Lehren der Technologie* [Discussion of the most important doctrines of technology] (Mohr und Zimmer, 1807), vii.

154 Johann Heinrich Moritz von Poppe, *Lehrbuch der allgemeinen Technologie* [Instruction book of general technology] (Friedrich Wilmans, 1809), 8–9. An instruction manual in "special technology" followed in 1819, also without deviating from Beckmann's approach.

155 Sigismund Friedrich Hermstädt, *Grundriss der Technologie oder Anleitung zur rationellen Kenntniss und Beurtheilung derjenigen Künste, Fabriken, Manufakturen und Handwerke, welche mit der Landwirthschaft, so wie der Kameral- und Policeywissenschaft in nächster Verbindung stehen* [Basic outline of technology or guide to the rational knowledge and judgment of the arts, fabricks, manufactures, and handcrafts, which stand in close proximity to agriculture, as well as to cameralism and police science] (Friedrich Maurer, 1814), 3.

than "a mere, often too short explanation of individual operations" and "superficial concepts."[156] Karmarsch, for his part, relies upon mechanical theory to take what could arguably be called a more metaphysical approach to technological processes than has been seen before, beginning with the categories of space and time. "As long as a body maintains its position, unchanged, we say it is in *rest. Motion* is what we call the change of this location, or the transition of the body from one location into the other. No motion succeeds without an inciting cause … That which can bring movement to bodies is called the *motive force*."[157] Karmarsch – who does not refer to Beckmann, Lamprecht, or Poppe – derives his mechanical doctrines of technology out of a network of basic concepts: rest, motion, and force. In a work published a decade later, the *Grundriss der mechanischen Technologie* [Basic plan of mechanical technology] of 1837, Karmarsch explores other aspects of technology. In that work he acknowledges that technological instruction can either be historically oriented (which he finds superficial) or have as its goal "a thorough, *practical* knowledge, albeit one that is enlightened and made clear by theory."[158] He also has a different approach to the concept of "special technology," which in his discussion starts to be used as a blanket term for earlier approaches ("the oldest and most common way of representing this science").[159] By contrast, the idea of a "general technology," relabeled as a "comparative technology" (*vergleichende Technologie*), focuses on the "means … *in themselves*" (*Mittel … an sich*), whereby these are understood as both material and immaterial entities: tools and machines, but also types of procedures (*Verfahrensarten*).[160]

There is further evidence that "technology" undergoes a process of redefinition in the German-language discussions of the nineteenth century in the work of Wilhelm Amandus Rüst and Johann Rudolf von Wagner. Rüst, a *Privatdocent* in Berlin, notably defines *Industrie* as that which "comprises every activity of man which is for the procurement or particular change of nature's products," – a definition that repeats almost verbatim earlier definitions of *Technologie*

156 Georg Altmütter, introduction to Karl Karmarsch, *Einleitung in die mechanischen Lehren der Technologie* [Introduction to the mechanical theories of technology]. (J. B. Wallishausser, 1825), v.

157 Karmarsch, "Introduction," in *Grundriß der mechanischen Technologie* [Basic outline of mechanical technology] (Im Verlage der Helwingschen Hof-Buchhandlung, 1837), 1.

158 Karl Karmarsch, *Grundriß*, vii.

159 Karmarsch, *Grundriß*, 3. Beckmann would likely quibble with this understanding of "special technology."

160 Karmarsch, *Grundriß*, 3.

itself around 1800.[161] Rüst also uses the increasingly common word *Technik* to refer to technology's subject matter when he writes that "technology therefore discusses what is practically achieved by technics [*in der Technik*]. It is essentially the theory of these practical achievements."[162] Wagner, for his part, also seems to bury technology within discussions of industry when he equates technology with "the theory of commercial industry [*Gewerbewesens*]" and further defines it as "the melting of science and technics."[163] He also makes a point of weakening the definition of technology even further by stating that it is "no independent science with its own principles, but rather only an application of the basic propositions and experiences of chemistry, physics, and mechanics."[164]

As a final thought, it is worth keeping in mind that the examples cited above may be indicative of trends in mid-century discussions of *Technologie*, but one would need a much broader investigation to achieve clarity on the many paths this particular problem word takes at that time. At the very least, one should also remember that there is no single path that technology follows. For every Langsdorf who emphasizes the materiality of *Technologie*'s domain of interest, there is a Karmarsch who connects it to the essential components of motion, space, and time. What is more, in the midpoint of the nineteenth century, we can still find writers such as Friedrich Knapp who both insist on the *logos* of *Technologie* and continue to speculate as to how different conceptual categories (such as the empirical, the rational, the general, and the special) might capture, in theoretical terms, the various kinds of reflections and practices that inform the sphere of technological thinking.[165]

---

161 W. A. Rüst, *Grundriss der Technologie: oder wissenschaftliche Beschreibung, Erörterung und Untersuchung aller technischen Verfahrungsarten, welche die Grundlage der ganzen Industrie bilden* [Basic outline of technology: or the scientific description, discussion, and investigation of all kinds of technical procedures which form the foundation of all industry] (Albert Förstner, 1844), 3.

162 Rüst, *Grundriß*, 4.

163 Johann Rudolf von Wagner, *Theorie und Praxis der Gewerbe. Hand und Lehrbuch der Technologie* [Theory and practice of industry. Handbook and instruction manual of technology] (Otto Wigand: 1858), 3.

164 Wagner, *Theorie und Praxis*, 3.

165 See Friedrich Knapp, *Lehrbuch der chemischen Technologie: zum Unterricht und Selbststudium* [Instruction manual of chemical technology: for instruction and self-teaching] (Friedrich Vieweg und Sohn, 1847), 2.

PART 1

From technologia *to* Technologie

ESSAY

Overview of the Entry on "Technologie" from Johann Heinrich Zedler's *Universal Lexicon*

The Latin word *technologia* had made its way into German as *Technologie* well before Johann Beckmann arrived on the scene.[1] One can find mention of it in various reference works, where it is usually defined as the doctrine of technical words (*Lehre von den Kunst-Wörtern*). Zedler's massive "universal lexicon," a sixty-eight volume work published between 1731 and 1754, is unique for providing a more extensive explanation of *Technologie*, one which includes some of the conceptual baggage accompanying this term's arrival into the German language. Zedler's entry includes a reflection on why one requires technical terms in the first place and makes a distinction between "unscholarly" and "scholarly" purposes, whereby the former would include the handcrafts and the latter philosophy or disciplines requiring erudition. Zedler also makes a distinction, where the scholarly usage is concerned, between "general" and "specific" terms: those used by all scholars (located in philosophy) and those unique to particular disciplines, such as theology. One of the things Beckmann, Lamprecht, Poppe, and others do around 1800 is to take the same distinction between the general and the specific and to repurpose it to define different aspects of technology.

In Zedler's entry on technology there is no hint, yet, of the further work that Johann Beckmann and others will accomplish to expand the role of *Technologie* in German language and culture. One will, however, notice a sensitivity towards language already evident in Leibniz's essay, "Unpresumptuous Thoughts Having to Do With the Use and Improvement of the German Language" (1697), discussed in the introduction to this volume. Like Leibniz before him – and like Beckmann after him – Zedler is concerned about the potential of technical terms to cause confusion. He recommends that one "keep such terms in the language in which they are usually found, and not translate them into the mother tongue without the approval of others" because "this causes obscurity."[2] However much the word *Technologie* evolves and adapts

1 Johann Heinrich Zedler, *Grosses vollständiges Universal-Lexikon aller Wissenschaften und* Künste [Great complete universal-lexicon of all sciences and arts], vol. 42 (Johann Heinrich Zedler, 1744), columns 508–509.

2 *Universal Lexicon*, vol. 42, column 509, s.v. "Technologie."

 | DOI:10.1163/9789004549548_003

over the course of the eighteenth and nineteenth centuries, one thing, at least, is clear amidst all the obscurity: the discourse of technology is marked by a particularly recursive gesture. It is a word that continuously provokes further reflections on the use of language itself.

TRANSLATION

Zedler's Lexicon Entry, "Technologie"

Technology, the doctrine of technical terms (*Kunst-Wörter-Lehre*), Latin *technologia*, is the name of the doctrine of technical terms, whereby those words are generally understood to denote things that belong to persons who find themselves in a particular circumstance. Because man should articulate his thoughts through words, those words either indicate things that could occur daily in the body politic and in all circumstances, or things belonging to persons living in a particular circumstance. The latter are called technical terms, because this situation would bring with it a particular skill or erudition. See Thomasius' *Introduction to the Doctrine of Reason*, chapter 4, §1.[1] One can divide technical terms into the unscholarly and scholarly, whereby the former occur in the handcrafts and common arts, the latter in the scholarly arts, and are in turn either general or specific. The general technical terms, which are used by all scholars, are located in philosophy, particularly in the so-called ontology, for example, *causa, effectus, principium, principiatum, subjectum, adjunctum, substantia, accidens*, and the like;[2] the specific ones are, however, located in theology, the study of law, and medicine. Whoever wants to deal with a science should understand such words, not that he would get to the truth through them, but rather that he would learn to understand others when they use this technical language. Scholars are particularly bound to the Latin language, regardless of the fact that such words also come out of other languages, such as Greek. One therefore does best to keep such words in the language in which they are usually found, and not translate them into the mother tongue without the approval of others. For this causes obscurity. There is a long tradition of using such words. Pythagoras attributed a particular meaning to the words *monas, tetraktys*, Plato *on, logos, nous, agathon*, just as Aristoteles does to *categoria, hypokeimenon, protasis*, and so forth.[3] Thus the ancient Greeks

1 Christian Thomasius (1655–1728), German philosopher. *Einleitung zu der Vernunfft-Lehre* (Introduction to the doctrine of reason). Halle: Christoph Salfeld, 1691.

2 These are common terms in Aristotle's ontological writings and could be translated as: *cause, effect, principle, principled thing* (which originates from the principle), *subject, adjunct, substance, accident*. I am using the translation for *principiatum* found in Edouard Hugon, *Cosmology*, trans. Romero Carrasquillo (Neuenkirchen-Seelscheid, Germany: editions scholasticae, 2013), 140.

3 The usual diacritical marks are omitted in this entry. Transliterated, these words are, in order: *monas* (from the Greek word for "unit," used by Pythagoras and later philosophers to refer

 | DOI:10.1163/9789004549548_004

already gathered such words, just as Timäus, Boethus, Harpocration, and others have done.[4] See Ménage's *ad Diogenes Laertius*, Book III, Section 63,[5] or Fabricius's *Bibliotheca Graeca* Book IV, chapter 36, § 10,[6] or Eckhard in the preface to *technica sacra*.[7] Rüdiger held a dissertation at Leipzig in 1700 titled *de usu & abusu terminorum technicorum in philosophia* [On the use and abuse of technical terms in philosophy], wherein, however, he understands *terminum technicum* to be such a word which indicates a thing according to its essence, parts, and qualities, for example, man, insofar as this word means a living, rational creature.[8] Walch's *Philosophical Lexicon*.[9]

---

to a single source or origin); *tetraktys* (triangular figure of ten points arranged in four rows, of mystical significance for followers of Pythagoras); *on* (the Greek word for being); *logos* (the Greek concept associated with words, reason or natural order); *nous* (the Greek concept associated with the mind or intellect); *agathon* (the Greek concept referring to the goodness or virtue of humans); *categoria* (the Greek logical term for a predication or category); *hypokeimenon* (the Greek concept referring to the underlying thing, also a material substratum); *protasis* (the Greek term for premise in Aristotelian logic).

4 It is not entirely clear which Timäus Zedler is referring to in this context. It could well be Timaeus the Sophist, who is thought to have written a lexicon of Platonic terms. Boethus of Sidon (*c.*75–10 BCE) was a Peripatic philosopher. Valerius Harpocration was a Greek grammarian active in the 2nd century CE.

5 Gilles Ménage (1613–1692), French scholar. *In Diogenem Laertium Aegidii Menagii Observationes & Emendationes*. Amsterdam: Henry Westen, 1692.

6 Johann Albert Fabricius, *Bibliotheca Graeca*. Hamburg, Christian Liebezeit, 1705.

7 Tobias Eckhard (1662–1737), German theologian and pedagogue. *Technica sacra, sive: De origine et usu verborum atque phrasium, quae sensibus mysticis a spiritu s. et ecclesia accomadata sunt.* (Sacred technics, or: on the origin and use of words and phrases whose mystical senses are adapted to the spiritual sense and the church.) Quedlinburg: Gottlob Ernst Strunz, 1716.

8 Andreas Rüdiger (1673–1731) was a German philosopher.

9 Johann Georg Walch (1693–1775), *Philosophisches Lexicon* (Friedrich Gleditsch, 1726).

ESSAY

Johann Beckmann (1739–1811): Technological Objects, Practices, and Ideas

1 Introduction

If one were to sketch out a geographical map of Johann Beckmann's life, it would extend in multiple directions from the small town of Hoya, in the province of Hannover, where he was born.[1] To the south, the map would contain the university town of Göttingen, where Beckmann matriculated in 1759 and later returned as a professor of philosophy in 1766. To the west, the map would need to include the Dutch cities of Amsterdam, Delft, and Rotterdam, where Beckmann toured both mines and other industrial sites, as well as visiting the universities in 1762. On the far eastern side of the map one would find St. Petersburg, where Beckmann taught mathematics, physics, and natural history at a German gymnasium. And to the north there is Sweden, where Beckmann spent time visiting the house and garden of the famed botanist Carl Linnaeus, who wrote a glowing *testimonium* to Beckmann's intellectual prowess and whose taxonomic system was important for Beckmann's own thinking about technological objects.

However clear the various details of Johann Beckmann's life may be when it comes to his travels, both within German territories and to points beyond, the intellectual terrain of his life is, by contrast, a matter of some dispute. Depending on whom one asks, Beckmann is either significant for being the founder of modern technology, or a mere curiosity item that had already become antiquated by the end of the nineteenth century. The purpose of the present essay and the translations of this volume is to offer an additional perspective to help readers determine for themselves the value of Beckmann's contribution to the conceptual history of technology. It seeks to demonstrate and interpret what Beckmann accomplished, rather than what is missing. With this objective in mind, the first section of the essay discusses the degree to which the German word *Technologie* was already part of Beckmann's vocabulary prior to the publication of the 1777 *Guide to Technology*, and what it meant for him. The second

1 The biographical details in this paragraph were drawn from Manfred Beckert's biography of Johann Beckmann: *Johann Beckmann* (Teubner, 1983).

 | DOI:10.1163/9789004549548_005

section then gives a brief overview of the structure of the treatise before taking a closer look at the discursive practices of the preface and the introduction to the guide.[2] It will show how Beckmann's concern for the observation and ordering of technological objects, and the definition of technical terms, is connected to more far-reaching questions about the basis of technology.

2 *Technologie* in Beckmann's Vocabulary

Beckmann was thinking about technology – in a particular sense of this word that will be described in more detail below – well before the publication of his 1777 guide. In order to understand the nuances of how Beckmann uses the term "technology" in his guide, one therefore first needs to consult his earlier publications. To begin with, one can compare the first two editions of his *Grundsätze der teutschen Landwirthschaft* [Principles of German agriculture], from 1769 and 1775. In a list that contains brief descriptions of various sciences that deal with the products of nature (such as the history and theory of nature, as well as agriculture, etc.), one finds in each edition a reference to the science that processes these products for human consumption:

> Die *Handwerkswissenschaft* lehret sie verarbeiten und verädeln.[3] (1769 edition)
> (The *science of handcrafts* teaches the processing and refinement of [natural products])
>
> Die Technologie lehret sie verabeiten und verädeln[4] (1775 edition)
> (Technology teaches the processing and refinement of [natural products])

2 There has been much written about various aspects of Beckmann's life, most of it in German, and nearly all of it from a purely historical perspective as opposed to one interested in a conceptual history. For German publications, the co-edited volume by Günter Bayerl and Jürgen Beckmann from 1999 provides an excellent overview of various phases of Beckmann's biography. Eric Schatzberg's *Technology: Critical History of a Concept* (2018) contains a short – and indeed, critical – section on Beckmann (see pages 77–81). For a different perspective, readers should consult Andre Wakefield's essay on "Beckmann in Göttingen," published in *La Technologie générale*. Wakefield encourages readers to challenge the Wikipedia version of Beckmann's biography and makes the argument that it was Beckmann's early work on agriculture, rather than the later writings on technology, that established his career.

3 Johann Beckmann, *Grundsätze der teutschen Landwirthschaft* [Principles of German Agriculture], (Johann Christian Dieterich, 1769), 3.

4 Johann Beckmann, *Grundsätze der teutschen Landwirthschaft* [Principles of German Agriculture], (Johann Christian Dieterich, 1775), 4.

Beckmann's use of the word *Technologie* is not an isolated occurrence in the 1775 edition, but rather one of twelve mentions that refer to, among other things, the "knowledge of technology"[5] and the "parts of technology,"[6] as well as more general musings as to "what belongs to technology."[7] In a separate publication, the *Physikalisch-ökonomische Bibliothek* [Physical-economical library], where Beckmann reviewed new publications in scientific fields of interest to him, *Technologie* is already used to refer to "knowledge of the handcrafts" in the third volume, published in 1772.[8] In the following volumes, all published before the 1777 *Guide to Technology*, Beckmann refers to teaching technology,[9] the lack of a dictionary of technology,[10] and the history of technology.[11]

Without knowing exactly what Beckmann's motivations were for introducing the word into his lexicon, it is clear that, as an item of interest and a technical term in its own right, technology gains more nuance during the years prior to 1777. First of all, it is an object of knowledge – one with a history. It is also complex enough to be divided into various subcategories. And finally, one of its tasks is to replace what Beckmann perceives to be the outdated term *Handwerkswissenschaft*, the science of the handcrafts. These various aspects of technology surely provided his readers a sense of what to expect from the *Guide to Technology*, but as the preface and introduction to this treatise will show, Beckmann continues to expand and develop his use of technology in unexpected ways.

3 The *Guide to Technology*

The first edition of the *Guide to Technology* contains a preface, an introduction (including, among other things, a list of 324 arts), and a table of contents listing

5 Beckmann, *Grundsätze* (1775), 5 and 336.

6 Beckmann, *Grundsätze* (1775), 6 and 457.

7 Beckmann, *Grundsätze* (1775), 508.

8 Johann Beckmann, *Physikalisch-ökonomische Bibliothek* [Physical-economical library], vol. 3.1 (Widow Vandenhoeck, 1772), 309.

9 Johann Beckmann, *Physikalisch-ökonomische Bibliothek* [Physical-economical library], vol. 4.1 (Widow Vandenhoeck, 1773), 274.

10 Beckmann, *Physikalisch-ökonomische Bibliothek* vol. 4.1, 215.

11 Johann Beckmann, *Physikalisch-ökonomische Bibliothek*, vol. 5.1 (Widow Vandenhoeck, 1774), 507. Annette Fröhner cites Beckmann's mention of a "technological article" (*technologischen Artikel*) in Krünitz's *Oekonomische Enzyklopädie* from as early as 1774. See "'Wenige, welche lesen mögen, werden nicht mit Nutzen und Vergnügen dieses Buch brauchen können.' Johann Beckmann und Johann Georg Krünitz: Reflexionen und Einflüsse," in *Johann Beckmann (1739–1811)*, ed. Günter Bayerl and Jürgen Beckmann (Waxmann, 1999), 175.

16 important arts, each of which is given its own chapter.[12] The preface argues its case for the importance of technology on two fronts. On the one hand, there is an economic and cultural argument to be made. The knowledge of handcrafts, 'fabricks' (the common English equivalent for *Fabrik* around 1800), and manufactures[13] is indispensable for future administrators. They need to know which industries thrive best in which geographical locations, how to procure the necessary resources, and how to tell good suggestions from bad or malicious ones – all in the name of contributing to the common good. To achieve these goals, one needs to keep an open mind. This includes being willing to learn from the successes of foreign countries (Beckmann recommends that educational tours should include industrial sites, not just museums and concert halls) and also re-evaluating traditional prejudices against the mechanical arts. To this end, Beckmann draws upon two important sources. One is Heraclitus's (reputed) statement that the gods also reside in humble abodes.[14] The other is the French *Encyclopédie*, where Beckmann cites an idea from Diderot's entry on "art": "In what physical or metaphysical system do we find more intelligence, discernment, and consistency than in the machines for drawing gold or making stockings?"[15]

Beckmann's readers have pondered whether, with the *Guide to Technology*, one is dealing with an "autonomous discipline" or an "auxiliary science," as Jochen Hoock frames the question.[16] In addition to the economic case to be made for the study of technology, and the desired re-evaluation of the status

---

12 The importance of this publication has been understood in various ways. Juergen Heinrich Maar has made a case that "modern chemical technology begins with the publication" of Beckmann's *Guide*.

See J. H. Maar, "Johann Beckmann (1739–1811) and Modern Chemical Technology," *Substantia* 7.1 (2002): 83.

13 Ulrich Troitzsch has observed that the definition of *Manufaktur* was vague at the end of the eighteenth century and was often described as having a similar "sphere of activity" to the fabricks or workshops (Troitzsch, "Manufakturen in Deutschland," 611).

14 The source of this anecdote is chapter five of Aristotle's treatise, *On the Parts of Animals*.

15 Denis Diderot (attributed), "Art," in *The Encyclopedia of Diderot & d'Alembert Collaborative Translation Project*. http://hdl.handle.net/2027/spo.did2222.0000.139 (accessed August 16, 2023).

16 Jochen Hoock acknowledges that technological discourse is one area where Beckmann seeks to "determine the theoretical place/position" of his first publications, but does not wish to overlook the fact that the opening lines of Beckmann's *Anleitung zur Handlungswissenschaft* [Guide to the science of trade] from 1789 "place technological discourse at the same level as practical arithmetic, the knowledge of wares, and applied geography." Jochen Hoock, "Johann Beckmann et la place du discours technologique dans la pensée économique de la deuxième moitié du xviiie siècle" [Johann Beckmann and the place of technological discourse in the economic thought of the second half of the

of handcrafts that should accompany it, the preface also lays the groundwork for Beckmann's concept of technology. It grants technology a mediating function between the pure or abstract sciences on the one hand and interactions with physical objects on the other. Acknowledging this mediating function is important, because it reflects dually on what technology is and what it does. Beckmann's preface suggests that when speaking about technology's theoretical potential, the physicality of the world of technological objects should not be downplayed, since these are objects that "arouse wonder" and demand as much "intellect, knowledge, reflection, acumen, and judgment as any scholarly problem is able to do."[17] Some years before Goethe encourages those who study plants to view them with the "eyes of the body" and the "eyes of the mind" in productive conjunction,[18] Beckmann proposes something analogous, a kind of doubled vision that is both analytical and empirical, able to navigate between the physical object and a more abstract kind of seeing.

In the introduction to the *Guide to Technology*, Beckmann begins by taking a few steps back from where the preface left off in order to give his readers the big picture of technology's cultural significance and what kind of knowledge it entails. He only arrives at a definition of technology in the twelfth paragraph: what comes before is a series of carefully crafted rhetorical gestures, including both an introductory and a final epigraph. The first epigraph stems from the Italian writer, Abbé Anton Genovesi (1713–1769), whose *Lezioni di Commercio o sia d'Economia Civile* (Lessons in commerce or civil economy, 1764) was translated into German as the *Grundsätze der bürgerliche Oekonomie* (Principles of the Bourgeois Economy). Genovesi defined the pinnacle of culture as the "blooming" of the technological arts, along with achievements in literature and science, because together they not only expand mankind's understanding, they also "enlighten him, and show him in the most insignificant trades that which he would otherwise not see."[19] Beckmann repurposes Genovesi's footnote to

18th century], in *La technologie générale*, ed. Guillaume Carnino, Liliane Hilaire-Pérez, and Jochen Hoock, trans. Joost Mertens (Presses Universitaires, 2017), 106.

17 P. 60.

18 Goethe writes that "there is a difference between seeing and seeing, that the eyes of the mind have to operate in a constant living connection with the eyes of the body, because otherwise one runs the risk of seeing and yet overlooking." Johann Wolfgang Goethe, "Entdeckungen eines trefflichen Vorarbeiters" [Discovery of an excellent predecessor] in *Sämtliche Werk*, vol. 12, *Zur Naturwissenschaft überhaupt, besonders zur Morphologie* [On the study of nature in general, particularly on morphology], ed. Hans J. Becker, Gerhard H. Müller, John Neubauer, and Peter Schmidt (Hanser Verlag, 1989), 85.

19 Abbé Genovesi, *Delle Lezioni di Commercio o sia d'Economia Civile* [Lessons in commerce or civil economy] (Federico Agnelli, 1768), 80.

this statement as the epigraph to the *Guide to Technology*: "Every art, however insignificant it may be, has its principles, and its mechanism, which only the philosopher can perceive. Thus the theories of the lowest arts may be brought into a science."[20] The epigraph therefore offers a model of inductive reasoning applicable to a technological context, whereby the right kind of philosophical gaze can infer general principles from the practice of technological arts.

The first paragraphs of the introduction lead up to a definition of technology by answering basic questions. These include, what is the difference between a "raw" and processed natural object? What is a handcraft, and a craftsman? What is an art and what is a trade? What are the stages by which one advances from apprentice onward? These definitions are also useful to present-day readers who might be unsure what the difference is between "locked" and "unlocked" handcrafts (in the case of the former, the number of practitioners in a particular location is regulated), or between a manufacture and a fabrick (no "fire and hammer" is used in the case of the former, unlike the latter).

The explanations continue through the first nine paragraphs and gradually home in on the physical context of production, with definitions of "material," "hand tool," "workshop," "wares" and, finally, the phrase "technical term" itself. This last item on this list provides Beckmann with a transition from the physical context of technological objects to reflections about language, which is the focus of paragraph nine. Like Leibniz, Beckmann maintains that technical terms for the handcrafts comprise an as yet unacknowledged wealth of language, and he adds that there is also a problem with an excess of technical terminology. The overabundance of synonyms for various technical objects is a hindrance to working, in Beckmann's words, "philosophically" or "systematically."

The final three paragraphs of the introduction are the most important for understanding how Beckmann positions the concept of technology in this treatise. They define technology (§12), clarify Beckmann's own theoretical vantagepoint (§13), and offer Beckmann's vision of "natural orders" (§14). In §12, Beckmann offers a straightforward definition of technology as "the science which teaches the refinement of the natural products, or the knowledge of handcrafts."[21] The "or" clause should not be read here in the sense of offering two distinct alternatives, but rather as a restating or reframing of the same.

---

20 Genovesi, *Delle Lezioni di Commercio*, 80. The quote from Genovesi has a resonance in the late-eighteenth century German technological literature that extends beyond Beckmann's treatise. One can also find it in Benedikt Hermann's *Über die Einführung des Studiums der Technologie* [On the introduction to the study of technology], from 1781.

21 P. 75.

The distinction is important because it speaks to the dual nature of technology as concerned with objects (here: the "refinement" of them) and our reflections upon them (i.e., the "knowledge" of the handcrafts that shape these very objects). Readers familiar with the preface to Beckmann's *Guide to Technology* may recall that there he already sets up a relationship between the activities of the workshop and our understanding of it – between practice and observation. In the preface, he writes,

> Whoever wishes to visit the workshops with these sheets [i.e., with the *Guide*] in hand, for him they will – at least I dearly wish and hope so – facilitate his viewing of the works in proper order, that he not overlook any, that he understand the language of the workers, that he more quickly comprehend the mechanism of the tools, and in general that he see more things and better understand them than if he had seen them without this little aid.[22]

In the introduction, however, Beckmann frames the relationship between reading about the arts and their actual observation in a somewhat different way. Immediately following his definition of technology, he suggests that it has the ability to offer something that cannot be learned from either direct observation or by listening to a master craftsman:

> Instead of just being shown in the workshops how one should follow the guidelines and customs of the master for the fabrication of wares, technology gives, in systematic order, thorough instruction [*Anleitung*] as to how one finds the means from true principles and reliable experiences for precisely this goal, and how one should explain and use them for the processing of the phenomena at hand.[23]

There is some irony to be discerned in the fact that, in what is arguably the central passage of the *Guide to Technology*, Beckmann defines technology by *its* ability to guide and instruct us. At the same time, he insists upon the dual nature of technology as something whose efficacy stems both from "true principles" (*wahre Grundsätze*) and "reliable experiences" (*zuverläßigen Erfahrungen*). Beckmann therefore seems to take Genovesi's idea that every art has "principles" and a "mechanism" only perceptible to a philosopher one step further by repurposing such philosophical gain for further use in the workshops.

22 P. 62.
23 P. 75.

Just as he did in the preface, in the introduction Beckmann draws upon the French encyclopedia project, in this case with two quotes. The first, a lengthy excerpt from Diderot's letter to Guillaume-François Berthier, argues for an equal place of the mechanical arts alongside the liberal arts. The second is of more strategic importance, placed as it is to comprise the entire first supporting paragraph after Beckmann's definition of technology. This second excerpt from the French encyclopedia project comes from the entry on "art" and begins by stating that the same dual nature with which Beckmann constructs his idea of technology (the refinement of objects, coupled with our knowledge of them) is, in fact, innate to the arts themselves:

> it is evident that every art has its speculative and its practical aspect: the former consists in knowing the principles of an art, without their being applied, the latter in their habitual and unthinking application.[24]

Beckmann and Diderot share the opinion that the manual dexterity required for handcrafts does not tend to accompany the ability to reason. Whether or not one agrees with this point of view, it speaks to Beckmann's conviction that technology falls on the side of speculation and can therefore, to a degree yet to be determined, be theoretical in nature.

With the distinction between speculation and practice in mind, Beckmann's following statement about his use of the word *Technologie* gains some additional contour. If historians of technology know of any quote from Beckmann's *Guide*, it is likely to be the one where he writes that he has "dared to use *technology* (*Technologie*) instead of the designation history of art (*Kunstgeschichte*), customary for some time, which is at least as incorrect as the designation *natural history* is for the study of nature."[25] One can look at this claim from multiple perspectives. With the recently stated distinction between speculation and practice in mind, the move away from a history would appear to be analogous to the move from empirical facts to a reflection on them; the "logos" of *Technologie* would support this point, as would the working definitions Beckmann provides of each term. In this reading, *Kunstgeschichte* is "the story

24 Denis Diderot (attributed), "Art," in *The Encyclopedia of Diderot & d'Alembert Collaborative Translation Project*. Trans. Nelly S. Hoyt and Thomas Cassirer (Michigan Publishing, University of Michigan Library, 2003), http://hdl.handle.net/2027/spo.did2222.0000.139 (accessed August 16, 2023). Originally published as "Art," *Encyclopédie ou Dictionnaire raisonné des sciences, des arts et des métiers*, 1:713–717 (Paris, 1751).

25 P. 75.

(*Erzählung*) of the invention, the development, and the further fate of an art or handcraft" and *Technologie* "explains the results and principles of all works."[26]

What Beckmann means when he says that he has "dared" to use the word *Technologie* is not precisely clear, however. Should the sentence be read as a self-referential statement that points to this particular passage in his introduction as a kind of turning point? Or does it gesture back to what Beckmann has been doing in his writing and his pedagogical practices, given that he has already been teaching courses on *Technologie* at Göttingen? Taking a more critical stance, one could also wonder why Beckmann claims to be using *Technologie* as a replacement for *Kunstgeschichte* (as a history of the arts) when just two years prior, in the *Principles of German Agriculture*, the term *Technologie* was clearly a substitute for *Handwerkswissenschaft* (the science of handcrafts). One take-away is that *Technologie*, as Beckmann understands it, needs to be a term pliable enough to encompass both the historical and the scientific aspects of handcrafts.

Beckmann concludes paragraph 12 with an analogy between *materia medica* and what he proposes to refer to as *materia technologica*, a comparison which allows him to transition into the question of how the various handcrafts should be ordered. In the eighteenth century, the term *materia medica* comprehended, according to a source from the time period, "all those bodies which are either used in Medicine in their natural state, or afford Preparations that are so," such that "These Bodies belong partly to the *fossile*, partly to the vegetable, and partly to the *animal Kingdom*."[27] It stands to reason that the empirical aspect of Beckmann's technology would require the same: it was necessary both to have an adequate inventory of the various natural products that became technological objects and to find ways of ordering the handcrafts that require them. Paragraph 12 describes various ways of dividing the handcrafts, including by material (stone, metals, etc.), by use (for nourishment, clothing,

---

26 P. 76.

27 John Hill, *A History of the Materia Medica* (T. Longman, C. Hitch and L. Hawes, 1751), 1. Stéphane Schmitt, in the context of a study of eighteenth-century encyclopedias and dictionaries focused on natural history, has also noted that "the tradition of alphabetical works was richer, more ancient, and more diverse in botany than in other scientific fields" and that already in the sixteenth century, "physician and naturalists had published a number of alphabetical lists, indexes of synonyms, and catalogs of plants and their properties, all of which were rendered necessary by the medical use of vegetal substances and the importance of identifying species." Stéphane Schmitt, "Jacques-Christophe Valmont de Bomare's *Dictionnaire raisonné universal d'histoire naturelle* (1764)," in Jeff Loveland and Stéphane Schmitt, *Specialized Dictionaries and Encyclopedias, 1650–1800* (Liverpool University Press, 2024), 204.

decoration, etc.), and by guild status (handcrafts that were either guilded or free). For Beckmann, however, "all these divisions do not serve as the basis for technology."[28] He rejects them wholesale, claiming that it is instead "most advantageous to bring into one division the handcrafts whose greatest works have an equivalence or similarity in the process itself and in the principles whereupon they rest."[29] In other words, Beckmann prefers to focus on similarities in processes of making which would then take precedence over any differences in material, function, or cultural status – it is Bacon's idea of translation retooled for an eighteenth-century context.[30] Beckmann is the first to admit that his proposal is not watertight,[31] particularly given that some handcrafts are a composite of very different kinds of work. He acknowledges that "this division also has its difficulties" but concludes, with a hint of defiance, "it certainly has fewer than all the ones known of before."[32]

The fourteenth, and final, paragraph of the introduction is simply a list of what Beckmann refers to as the "natural orders of the handcrafts and arts," followed by a final epigraph. The list contains 324 individual items, beginning with "*Handwerk der Schlächter*" (Handcraft of the slaughterers) and ending with "*Bildgiesserey. Statuen*" (Statue casting. Statues). It is divided into 51 sections, headed by Roman numerals. Although Beckmann does not give a heading for each section, it is often not difficult to follow his train of thought. Section XXIV, for example, contains arts relating to paper, from the art of making it to book binding. In other cases, however, it can be more difficult for modern readers to discern the common denominator. Section XXV begins with the preparation of pencils but also includes the making of such diverse items as sealing wax

28 P. 76.

29 P. 76.

30 One could also consider the problem of technological "translation" from a somewhat different perspective, in the context of the "transfer" or "diffusion" of knowledge. For David Garrioch, who has written about artisan mobility in the European context, this idea should come with the caveat that "ironically, an emphasis on 'transfer' or 'diffusion' … downplays the role of artisans as innovators, portraying them primarily as intermediaries rather than as active participants in the creation of skills" (Garrioch 2022, 16).

31 Like Diderot and D'Alembert before him, Beckmann is well aware of the "arbitrariness of epistemological models," and that their "usefulness to the knowing subject" should be the criterion, rather than a "transcendent order" (Brewer, "The *Encylopédie*: Innovation and Legacy," 54). This was a sharp break with tradition, an attempt "to supplant both the cosmic systems and canonical principles of medieval theologians, and the dialectic of ancient and modern traditions adopted by Renaissance humanists, with models inspired by the work of engineers, geometricians, and empirical scientists" (Roche, "Encyclopedias and the Diffusion of Knowledge," 173).

32 P. 77.

and artificial flowers. Perhaps with the diversity of the arts in mind, Beckmann chooses to conclude the introduction with a reaffirmation of the interconnectedness of the handcrafts and the arts, using as his final quote a line from Tertullian's treatise on idolatry: *Nulla ars non alterius artis aut mater propinqua est.* (There is no art which is not either the mother or relative of another.)

TRANSLATION

Beckmann's Preface to the *Guide to Technology*, 1st Edition (1777)

The knowledge of handcrafts,[1] fabricks,[2] and manufactures is indispensable to anyone who wishes to dedicate himself to civil government and cameralist sciences. For regardless of what one might wish to organize, construct, arrange, evaluate, govern, maintain, improve, and use, one will at the very least be required to have knowledge of it. Regarding questions such as which industries are lacking in our fatherland, which of the missing ones could be advantageously introduced, from where one should take materials for that purpose, from where one should procure artisans for it, where the most suitable location is to which one can direct them, what is holding back the handcrafts that we have, how can they be helped, how much each contributes to the common good, and how one can calculate their profit, cameralists will only be able to answer these and many other important questions once they have acquired that knowledge. Where that knowledge is lacking, charlatans will be honored, and ignorant, deceiving swindlers (*Projecteurs*) will continue to be accepted and rewarded until finally, due to an unseemly desperation, and due to reprehensible mistrust of their own judgement, people reject both the good and the bad suggestions at the same time, leaving the former (good) suggestions to the more intelligent English, and leaving the latter (bad) ones to the individuals. Where this knowledge is lacking, industries will be abandoned to their own fates, or, what is even worse, they will receive regulations and laws that they cannot follow, or by which they will be completely ruined. Then offices of inspection will emerge that hold back diligent labor and choke industry. You strain out a gnat but swallow a camel.[3] In France and England,

1 *Anleitung zur Technologie, oder zur Kenntniß der Handwerke, Fabriken und Manufacturen, vornehmlich derer, die mit der Landwirthschaft, Polizey und Cameralwissenschaft in nächster Verbindung stehn. Nebst Beyträgen zur Kunstgeschichte* [Guide to technology, or to the knowledge of handcrafts, fabricks, and manufactures, primarily those which are most closely connected to agriculture, policy, and cameralist science. Alongside contributions to the history of the arts] (Göttingen: Verlag der Wittwe Vandenhoeck, 1777).

2 *Fabrik* is usually translated as "factory" in modern German. Instead, I have used "fabrick" throughout, which is the term used in English around 1800.

3 Matthew 23:24.

 | DOI:10.1163/9789004549548_006

the manufacture-colleges think the same as the consistories[4] in Germany. The latter do not ask whether their candidates have studied pandects or therapeutics, but whether they have studied theology, and can prove as much; and the former only take on candidates who have had the honor of getting to know the manufactures, whereby they seek their bread.

The knowledge of handcrafts, fabricks, and manufactures is extremely useful to whomever wishes to occupy himself with agrarian production and trade. For the extraction of raw materials occurs with the intention of giving them to manual laborers, either directly or through merchants, and the knowledge of these products will be all the more advantageous, the more they are in the condition that the artisan demands, and in the condition that the agriculturalist, who is instructed about such requirements, can quite often achieve. Such a person knows to sort through his wares and to seek out, for every kind, the relevant merchant who can best use them, and who can also pay the most; whereby another hopes for the greater profit with rough wool, whose weight he deceitfully increases with dust and moisture, hates silk spinners,[5] and gives good and bad wool, just as it has been shorn, and wood, as it has been cut, to the retailer, who then, when he possesses the technological insight lacking in the agriculturalist, extracts the greater profit that the other could have had. When the processing of his products is left freely to the agriculturalist, he can only undertake it if he has knowledge of them, and then he can often increase his advantage numerous times. He profits as merchant and craftsman.

If those who wish to go on journeys were first to familiarize themselves with various industries, and acquire the desire and ability to investigate their condition among foreign lands, and to note their advantages and new inventions, they would return with still richer bounty than is usually the case, at the very least not with a loss, which is more common. Should this become the custom in our fatherland, then the German baron would bring back more from Paris than new fashions and folk songs. Then he would see more in Italy than what the guide shows everyone who pays him, more than the antiquities observed and described by so many. Then, in England, he would not only visit Covent-Garden, Drury Lane and Vauxhall, but also the workshops of his fellow countrymen, who help the English deserve pride of place before the Germans with regard to the arts. Then German thalers would certainly be exported, but

4 A term from Catholicism: a consistory was an ecclesiastical court.

5 A term for a sheep with unusually fine wool. See *Deutsches Wörterbuch von Jacob Grimm und Wilhelm Grimm*, digitalized version in the dictionary network of the Trier center for digital humanities, version 01/23, <https://www.woerterbuchnetz.de/DWB?lemid=S25076>, accessed on August 18, 2023, s.v. "Seidenspinner."

foreign knowledge would also be imported, and there would still be the question of who paid the account balance, the German or the Englishman.

For the actual scholar, who is neither agriculturalist, nor merchant, nor cameralist, knowledge of technology is no less important. Mathematicians and investigators of nature cannot elevate their sciences any higher than by applying them for the use of industries, whose improvement is also the immediate improvement of the state. Then they bridge the great distance between erudition and that which can be used in common life; namely the distance which the practical statesman believes to observe which, because he prefers what is the most indispensable, immediately useful, and frequently applicable to that whose absence he does not sense and whose application he seldom sees, at times commits a sin against sciences, or is their patron from other reasons than from conviction. Then the scholar will find, as if in a new world, objects in the workshops which will arouse wonder, however hardened he may be: objects that demand just as much intellect, knowledge, reflection, acumen for their judgement and explanation as any scholarly problem is able to do (*).[6] Here, too, reside the gods,[7] said Heraclitus, as ignorant people were astonished to encounter him at a smithy. Then learned men will help elevate trades, without which the state cannot exist, but which, because in Germany one has always, due to ignorance and prejudice, held them to be simplistic, unworthy

6 (*) (Translator's note: Beckmann's own footnotes are indicated both in his printed edition and the present translation with an asterisk. In Beckmann's text, the present note is in French, from the French *Encyclopédie* entry "art" usually attributed to Denis Diderot, although Beckmann lists *D'Alembert* as the author). The quote is: "they are well worth the trouble whether we think of the advantages they bring us or of the fact that they do honor to the human mind. In what physical or metaphysical system do we find more intelligence, discernment, and consistency than in the machines for drawing gold or making stockings, and in the frames of the braid-makers, the gauzemakers, the drapers, or the silk workers? What mathematical demonstration is more complicated than the mechanism of certain clocks or the different operations to which we submit the fiber of hemp or the chrysalis of the silkworm before obtaining a thread with which we can weave? What projection is more beautiful, more subtle, and more unusual than the projection of a design onto the threads of a simple and from there onto the threads of a warp? What can conceivably be more subtle than the art of shadowing velvet? I could never enumerate all the marvels that amaze anyone who looks at factories, unless his eyes are closed by prejudice or stupidity." Denis Diderot (attributed), "Art," in *The Encyclopedia of Diderot & d'Alembert Collaborative Translation Project*, trans. Nelly S. Hoyt and Thomas Cassirer (Michigan Publishing, University of Michigan Library, 2003), http://hdl.handle.net/2027/spo.did2222.0000.139 (accessed August 16, 2023). Originally published as "Art," *Encyclopédie ou Dictionnaire raisonné des sciences, des arts et des métiers*, 1:713–717 (Paris, 1751).

7 In Beckmann's text, this quote is printed in Greek. The source of this anecdote is from the fifth chapter Aristotle's treatise, *On the Parts of Animals*.

occupations, have sunk to the most common, ignorant, and impecunious class of commoners, where these trades, like seeds on rocky ground, certainly sprout but, due to a lack of sustenance and care, never fully mature.

Jurists, apart from those whose positions are related to finances and the polity,[8] who think to practice one day, will neither defend the rights of the handcrafts, nor contest (the one seldom occurs without the other), nor settle their disputes, if they do not know their labors. Ramazzini, Linné,[9] and others have already proven, through their examples, that the doctor, if one wants to distinguish him from the investigator of nature, can use knowledge of handcrafts. The theologian, – Mathesius[10] at least, whom Luther and Melanchthon honored as a knowledgeable spiritual curate, – knew the trade of those to whom he preached, and his prayers are still today, after two hundred and fifteen years, sought after, purchased, read, and used, while in the meantime a countless pile of homilies by empty ascetics has entirely perished.

I have only included this as something superfluous (*).[11] Because, to speak the truth, one must give the praise to our century, that everywhere at this time people are beginning to extract what is more useful and to agree how appropriate and to some degree necessary the knowledge of agriculture, the manual arts, and trade is to those who are required, or inclined, to serve the state. That means, in other words, that one finally no longer doubts that the science of which I speak deserves to be taught and learned. I should thus be permitted to say what I must, with regard to the present guide.

This guide is not intended to educate any weavers, brewers, or any handworkers at all, other than those who require many accomplishments and techniques (*Handgriffe*) for the practicing of their arts, which must all be acquired individually, through lengthy practice, but which are unnecessary for those whom I seek to serve. The captain must know the jobs of the artillery shooters, but it does not shame him if those men know how to mount a defense

8 *Polizeywesen* is defined as "the Polity, the Order concerning the Security and Conveniency of the Inhabitants in a City and also throughout the whole Province or Country" in John Ebers, *The New and Complete Dictionary of the German and English Languages*, vol. 2 Leipzig: Breitkopf and Haertel, 1798, s.v. "Polizeywesen."

9 Bernardino Ramazzini, Italian physician (1633–1714); Carl Linnaeus, Swedish botanist (1707–1778).

10 Johannes Mathesius, German minister (Lutheran reformer), (1504–1565).

11 (*) If I had wanted to present the usefulness of technological knowledge completely, then I would have to have said that it serves everyone, whoever he may be, educated or uneducated, by both the purchasing and the use of wares. Whoever understands their processing understands how to select, to estimate, and to use the best ones to their advantage, as well as to protect and improve them.

faster than he does. The agriculturalist has to understand the thresher, but he does not need the skills of threshing; he could also be lacking the bones and muscles for it. The handworkers are to the cameralists, as the fieldworkers are to the agriculturalist, as the apothecary is to the doctor. Thus one will not find it lacking that I have only mentioned difficult labors that require many techniques (*Handgriffe*) without touching upon them. On the other hand, I have taken the trouble of determining the raw materials and secondary materials, referring to the tools and equipment for explaining terminology, describing the various works in the order in which they occur, or according to which they can most easily be understood, giving their fundamentals, touching upon the differences in tools and labors, which are not generally known. Throughout, I have followed the motto of saying much with few words. Whether or not I have achieved that remains to be heard from others, but I can offer the assurance that some of the adjectives have cost me more time and reflection than the judgement of this little book will cost some of the readers. If a few paragraphs, especially such ones where complicated machines are described, do not create a complete image, then I ask that one consider that these sheets, which should not be made more expensive by an engraving, are intended for lecture, whereby everything is to be elaborated orally, partly in the auditorium, and partly in the workshops in proximity to the machines themselves, or with models or with drawings. Whoever wishes to visit the workshops with these sheets in hand, for him they will – at least I dearly wish and hope so – facilitate his viewing of the works in proper order, that he not overlook any, that he understand the language of the workers, that he more quickly comprehend the mechanism of the tools, and in general that he see more things and better understand them than if he had seen them without this little aid. Thus the botanist sees, when he compares a plant with a precise, if cursory description, more of it than he would have noticed without a description, and he not infrequently sees even more than the author of the description has! One has to have tried to learn about the fabricks and manufactures without any preparation and guidance in order to know how hard it is to observe the workers and inquire about what is essential, the order of operations, and the arrangement of the tools, when these workers are for the most part not used to thinking about their occupations, and have still less desire and ability to explain them, who will become impatient with the ignorant stranger who holds them up with questions and suggestions and who, out of ignorance, conceal precisely that which the scholar or genteel traveler inquire about, like an imponderable secret.

I have primarily, among the handcrafts that stand in close relation to agriculture, polity, and cameralist science, chosen those to which I can take my auditors here in Göttingen or close by. Among the latter I include the neighboring salt works as well as Münden, Kassel and the instructive Harz. Perhaps some reader might note the lack of a handcraft that could have been expected here, according to my plan, and finds one or the other which, in his opinion, could have been left untouched without harm. But since I could not discuss all of them, I never would have been able to avoid such reproaches, no matter which ones I chose. In addition, it is certain that a thorough knowledge of a few fabricks and manufactures serves as a very good introduction to the knowledge of all others. I mean, one will become more familiar with them than if one had no prior practice in such investigations at all. Where is then the school, wherein all parts of a science have been taught, such that nothing else is left to be learnt through the student's own industry?

Perhaps some would have expected me to have made the connection between the described handcrafts and agriculture, and the sciences of governance (*Polizeywissenschaft*)[12] and cameralism (*Cameralwissenschaft*), and to have shown, for example, what kind of misuses should be prohibited in each case, and what kind of laws could be given to each. Only, I know from experience that one achieves clarity and completeness in lecture when one treats each of these sciences individually and in order, so that one allows technology to follow agriculture, the science of trade to follow technology, the science of governance to follow the science of trade, and finally cameralist science. Agriculturalists, handworkers, and merchants see to their private interests in

12 *Polizey*, as understood in the eighteenth-century context, not particularly easy to translate. It can be helpful to refer to a contemporary definition by Johann Heinrich Gottlob Justi, from the preface to his 1756 work *Grundsätze der Polizeywissenschaft*, that captures some of the nuances: "*Polizei* concerns itself with nothing but the maintenance and increase of the entire property of the state through good internal organisation, lending the republic all inner power and strength of which it is capable according to its condition. To this end it seeks to cultivate the lands, improve the state of subsistence, and maintain discipline in the common weal" (quoted in Tribe, "Cameralism and the Sciences of the State," 541). In an earlier essay, Tribe has outlined confusion surrounding the concept of *Polizei* from our present-day perspective in great detail: it might be tempting to distinguish between cameralism as "principally concerned with the administration of a state," *Oeconomie* as connecting the "activity of administration materially to the objective of happiness," and *Polizei* as concerned "with the general condition of order prevailing in the state," but this "neat separation is however only effective on the basis of modern conceptions of economy and polity that are not present in the eighteenth century" (Tribe, "Cameralism and the Science of Government," 266). Gradually, *Polizei* began to refer more to matters of security. According to a late-nineteenth century lexicon, it was "in the broadest sense … the totality of state activities directed internally to assuring the security and welfare of the state and its members." Franz-Ludwig Knemeyer, "Polizei," trans. Keith Tribe, in *Economy and Society*, vol. 9: 189.

the course of their trades. The polity leads them for the good of the entire state, that is, it commands and also prohibits when the advantage of individual citizens is not to the advantage of the entire society, and also when society itself, out of shortsightedness, fails its own. Cameralist science has as its purpose the income of the regent, or the public coffers, out of which the expenses of the state must be sufficient, and since the income is proportional to the wealth of the people, this purpose will never be reached without reasonable civil governance. Also, the citizen should always seek only his own private interest! That is why he lives in the state and contributes his share to its expenses, in order to master a trade, though not the one that belongs to the generally greater good, and to be allowed to pursue the former without holding himself back out of concern for the latter. He should only obey when the authorities command, and they should understand and observe their duties. Then the state will be the most ingenious machine people have ever produced, within which a number of wheels and cogs, both great and small, interlock with one another.

In order to make the somewhat monotonous presentation more appealing with a few flourishes, I have mixed in that which has become known to me, often through tedious research and through the comparison of various reports, about the history of the arts treated here, and about the inventors and the time of the inventions. I have, since this was not my primary objective here, left out the references in an effort to be more concise (*).[13] The actual, so-called history of the arts has, up to now, only been studied by a few, and in addition to that, only by those who themselves did not know them, but I entirely remove all of the fine arts, which excellent history writers have preserved. Nevertheless, this history is not just pleasant, but also instructive:[14] and we Germans have more reason than all of our neighbors to investigate it, since the most numerous and most important discoveries have undeniably been made by our countrymen, however much other nations claim such things. More than seven Greek cities quarreled with each other for the honor of being the birthplace of a poet. All of Germany allows itself to be robbed of the honor and advantages of its many discoveries, whereby other peoples have become wealthy, powerful, and fortunate – those who make Germany weak and despise it. It is uncommonly pleasant for me to see that hope fulfilled which I have expressed on page 375.[15] I already possess the five first sheets of the fifth part of Herr von Murr's *Journal*

13 (*) I have made the attempt to treat this subject more precisely in my *Contributions to the History of Inventions*, of which the first volume is now in print.

14 "To delight and instruct," from Horace's *Ars Poetica*, usually refers to poetry.

15 On that page, Beckmann expresses the wish to one day read Murr's account of all the arts, including the fine arts.

of Art and Literature (*Journal zur Kunst und Litteratur*),[16] and therein I find with pleasure the beginning of a Nurembergian history of handcrafts, from the thirteenth century through the middle of the sixteenth. Since I only received the beginning of this very valuable essay just after the entire printing of my guide, I have not used it here, but I take it as my duty to at least mention it to my readers, and at the same time to thank Herr von Murr for the regard he was pleased to take of my work.

To those who cannot imagine, or cannot admit, that agriculture, technology, and the science of trade can usefully be taught at university, I assure you that I know the contrary to be true from twelve years of experience, and that I could name men who now hold offices where they need such knowledge and would not hesitate to affirm it. Of course, the merchant is only educated in the office, the handworker only in the workshops, but it is laughable to expect that the future cameralist or anyone else would spend a few years at all the trades he is taught, whose theories he has to know.

Whoever may notice any mistakes that I have made, despite all care (and I do not think it is difficult, rather, I trust that some journeymen, and even apprentices, could point out mistakes), they should know that I admit without hesitation that I have not spent apprenticeship years at any of the handcrafts discussed here. I certainly visited them, frequently and attentively, both in and beyond Germany, but did not have the convenience, during the writing execution of the guide, to return to all of them as often as I wanted, in order to observe and to expand upon the one or the other. From persons of moral standing I intend to accept improvements and additions without having to blush, and I will acknowledge and use both with gratitude. How, the person who seeks fame through slander and libel shows that he is not worthy of being nurtured in the state whose laws he despises. He spreads and renews his own disgrace and is too much reviled to even enjoy the base pleasure of doing real harm.

Göttingen, 12 March 1777.

16 The title is actually *Journal zur Kunstgeschichte und zur allgemeinen Litteratur*. Christoph Gottlieb von Murr (1733–1811) was a German scholar based in Nuremberg.

TRANSLATION

Introduction to the *Guide to Technology*, 1st Edition (1777)

> Ogn' arte per vile che fia ha I suoi principi e il suo meccanismo, che non può esser avvertito che dal' filosofo. E quindi è che le teorie dell' arti le più vili, si possono ridurre a scienza.
>
> Lezioni d'economia civile dell' Ab. GENOVESI. I p. 102

> Every art, no matter how humble, has its principles and its mechanism, which can only be felt by the philosopher. It follows that theories of even the most humble arts can be reduced to science.
>
> Lessons in civil economics by the Abbot GENOVESI, vol. 1, 102[1]

∵

§1 Few products of the earth (*Naturalien*)[2] are suitable when raw for all the various kinds of use which humans can and wish to make of them. Most have to be worked upon, or *refined*.

1. *Raw*[3] is what natural products are called just as they are acquired, when they still have not been worked upon or refined in any way.

§2 The art of refining raw natural products or those already worked upon is called a *handcraft*. The person who possesses this art and pursues it as a

1 Antonio Genovesi (1713–1716), Italian writer of philosophy and political economy. Thanks to Luciano Pomatto for the translation of this quote.

2 *Naturalien* is a common term in the technological literature around 1800. "Natural products" and "products of nature" are both common translations at the time, as is "products of the earth," which I have chosen because it can also convey the impression of being somewhat more abstract and process-oriented. Ebers vol. 2 (1798), s.v. "Naturalien."

3 It might sound odd to use the word "raw" to refer to a category of objects that far exceeds those intended for cooking, but Ebers' dictionary uses this adjective broadly and refers, for example, to "raw thread," "raw silk," and "raw hide," in addition to other items. Ebers, vol. 2 (1798), s.v. "roh."

 | DOI:10.1163/9789004549548_007

trade is called a *craftsman*. *Masters* are those who pursue a handcraft for their own profit and are permitted to instruct others.

1. *Art* is what every occupation is called which, according to certain guidelines or rules, is carried out with a skill acquired through practice. Every handcraft is an art, but not every art is a handcraft. Otherwise, billiards and whist would be handcrafts. But the clockmaker, the silk weaver, the image caster,[4] – the broom binder, the maker of mousetraps, are handcraftsmen. The former are with this designation just as little humbled as the naturalist who dishonors mankind when calling humans animals. And just as little as the broom binder gains in status through this general designation, so little does the mite become more important by occupying a place in the catalogue of the animals.
2. Thus have our ancestors understood these words. *To craft* (*werken*) meant as much as work and refine and has been preserved in various combinations: for example, workshop (*Werkstelle*), work table (*Werktisch*), fireworks (*Feuerwerk*), fireworker (*Feuerwerker*).[5]
3. Therefore, arts which do not concern themselves with the refinement of natural products also do not belong here, even if one has given them a craftsmanlike designation. I thus leave hunting (*Jagdkunst*), riding (*Reitkunst*) etc. among the general name of arts. Others can divide them and distinguish them with names.
4. *Trade* (Gewerbe) is every occupation that is pursued with the intention of gaining money from it. If I am not mistaken, all trades can be brought together under the following divisions:
   1. Agriculture.
      – Animal husbandry, crop farming.
   2. Mining.
      – Acquisition of minerals.
   3. Handcrafts.
   4. Trade.
   5. Arts.
      – Music, artillery, ship-building, etc.
   6. Sciences.
      – The study of nature, surgery, history, theology ––– mathematics
   7. Private services.
      – Servants, farm hands, maids.

4 A *Bildgießer* would pour molten metal or plaster into preformed molds to create images.
5 The person who puts on the fireworks display.

8. Public service.
   – Judges, army commanders, regent, teacher, etc.

5. Certainly sciences belong to the trades, at least in the present times. Their objects are like the noble metals which are often reworked – at times adulterated, at times refined again – and thereby employ a lot of people in work and service. The former [i.e., the sciences] lose nothing of their worth thereby, as little as gold ceases to be gold, whoever may process it or make it into an alloy. The sciences do not become handcrafts simply because they, like handcrafts, are trades. Music and the art of turning-lathes do not become parts of statecraft or the art of governance even if they have similarly crowned virtuosos. Those who speak against the number of books, or against writing books oneself, would not be able to procure another appropriate trade for those who live – and have to live – entirely or partially by this. Those who cry out against the world in which scholars and so-called geniuses have to work in order to live want for those able to work to be fed by the industry of others. But no! Pray and work is a general law.[6] Leibniz was a librarian in Hannover, and Newton was the chief mint master in London. Is it proven that they would have been more useful to the world if they, without employment, had lived off of wealthy canonries?

§3 When Germany had no other towns apart from Roman border fortresses, everyone earned and took care of themselves, or allowed what he needed to be earned and taken care of by bondmen. But when Henry the Fowler (Heinrich der Vogler)[7] constructed fortresses, he assigned to the inhabitants of the castle (*Burg*) – the castle dwellers (*Bürgern*) – handcrafts, commercial trading, and sciences as trades. Since this time there have arisen in Germany Roman-formed trade guilds (*Zünfte*), craft guilds (*Gilden*), and trade associations (*Innungen*), or societies approved of legally, through trade letters, which exclusively pursue certain trades, primarily handcrafts. Handcrafts which did not maintain guilds were thereby denied being named as such and denied the same rank.

1. Only the emergence and establishment of the guilds belongs here. The question of their value today belongs to the governance (*Polizey*) of the handcrafts, that is, to the part of the town-governance which instructs the way of regulating the town trades, and thus primarily the handcrafts, to the greatest good of the state. They (the guilds) seem admittedly in

6 The phrase *ora et labora* is associated with the order of Benedictine monks.

7 Heinrich der Vogler (876–936) was King of Saxony from (912–936).

their creation to have been as useful as they are now harmful. Already in the thirteenth century an attempt was made to do away with them, but one had to give it up after it had cost enough blood in Würzburg, Goslar, and other places. Even in the monarchy of despotic France, the economists have not been able to push through the removal of the guilds.

2. The certificates, bills of sale, letters, and money of the trade societies are kept in the master's trunk (*Meisterlade oder Lade*)[8] that is opened when the handcraft – that is, the guild – is gathered. From there comes the expression, when the trunk is open (*bey offener Lade*).

§4 The more elaborate the handcrafts became, the more practice, experience, and time they demanded for their learning. Thus one began to take young boys as students, or to enroll and apprentice boys who, in order to maintain the prestige of the guild, had to prove their freedom and honesty with a birth certificate. After the apprenticeship years they were released with a letter of apprenticeship, or declared to be journeymen (*Gesellen*).

1. Instead of the term journeyman (*Gesellen*), some handcrafts have still kept the word servant (*Knecht*), for example millers, bakers, cobblers, brewers, tanners. Either these handcrafts are the oldest guilds, which could, when they were established, not have any helpers other than bondmen or servants, or they are more recent than the other handcrafts, and are still pursued by serfs (*Unfreyen*), since the others were already occupations of the freeborn.
2. In order to slow down somewhat the transition from compulsory apprenticeship (*Lehrzwange*) to mastery, handcraftsmen distinguish between young journeymen (*Junggesellen*) and foremen (*Altgesellen*).

§5 In part to broaden knowledge, and in part to prevent the journeyman from establishing himself directly next to the master, the law was created that journeymen must travel. To facilitate traveling or *wandering*, the *gift* was introduced at some trade societies. In order to distinguish learned journeymen from frauds, a ceremony was adopted, the handcrafts-handshake introduced, and documentation demanded.

1. Hereby arises the difference between gifted and ungifted handcraftsmen.
2. The first journeys of the German handcraftsmen were to Italy,[9] where the arts were older and more established. Quite a few German technical

8 *Lade* is the word for trunk. The *Meisterlade* is the master's trunk.

9 Beckmann uses the word *Wälschland, an* older term for foreign lands, usually where Romance languages were spoken. Technically, it could refer to Italy or France.

terms thus come from the Italian, and presumably more useful knowledge was retrieved by the handcraftsmen from Italian than by the little masters from France.[10]

§6 In order to make journeymen into masters, and to procure work for traveling journeymen, an inquiry is made by the older ones.[11] Whoever wishes to become master must first prove his talent with a *masterpiece*.

1. For the attainment of mastery, a ceremony has been introduced – somewhat ridiculous, somewhat painful – which is borrowed in part from the Roman emancipation, in part from habits of the clergy, and in part from the schools of the old philosophers. This hazing is perhaps supposed to scare away journeymen and limit the number of masters, or offer the opportunity for a celebratory feast, or it should provide the guilds with some of the prestige of the religious orders.
2. Among masters, the distinction between young master, old master, and guild master has also been adopted.

§7 In a few places, some handcrafts are limited to a certain number of masters, but others are not. The former are called *locked*, the latter *unlocked*. A few towns have wanted to keep some handcrafts for themselves alone, but other handcrafts are to be found everywhere. The former are called barred handcrafts, the latter are free.

§8 In more recent times, people have tried to prioritize certain handcrafts by naming them fabricks, manufactures, or arts. The first two terms can be given to every handcraft where it is produced in quantity, but with the distinction that fabricks use fire and hammer, but manufactures do not. Arts refer to handcrafts that demand great natural abilities and much secondary knowledge, also those which process jewels, gold, and silver and are not contained by guilds. This priority is justified, but in a scientific presentation one is neither required nor able to observe it.

1. People have tried to remove some arts from this class and raise them above all others, namely: those which demand some erudition, particularly knowledge of history and mythology (*Fabellehre*); those which

10 "The appellation of petit-maître is given to a certain set of young men of the court, who take the lead of the rest, and affect a superiority by their free and bold behavior." H. J. P., "Petit-Maitres," in *The Monthly Mirror: Reflecting Man and Manners* 17 (May 1804), 296.

11 The verb *umschauen*, most often used in the general sense of looking around, is given a special meaning here.

preoccupy themselves more than others do with the imitation of nature; and those which make what is beautiful and pleasant, rather than what is useful. I mean the so-called fine arts (*schöne Künste*), for example, painting, sculpture, the art of carving, etc. Their praise is so exaggerated that for a time it was thought their knowledge was only worth the attention of great and educated men, until it finally seemed that one should first look within the state for what is indispensable and useful and then for what is beautiful, or at least for both with the same enthusiasm. The fine arts are the foster children of excess, and this excess comes from those trades which one used to despise. In Europe and Germany, the part of the world which for the longest time and most successfully pursued agriculture, mining, fabricks, manufactures, handcrafts, and trade, the fine arts were also the earliest to sprout and grow. The closer the rest of the world comes to them in the pursuit of the mentioned trades, the closer it comes to them in the fine arts as well. But that means the latter more than the former desire to pluck flowers before one has sown the seeds. According to the explanation given above, however, the fine arts belong to the handcrafts, and my respect for the latter is too great for me to place them a step lower with this confession, or to offend their admirers, to whom I also belong.

2.[12] When people examined the products of the arts, they realized that some were primarily created by the mind, others by the hands. This is *part* of the cause for the pre-eminence that some arts have been accorded over others, and of the distinction between liberal and mechanical arts. This distinction, although it is quite justified, has led to bad consequences because it has given a low name to people who are very worthy and useful, and encouraged us in a certain natural laziness. We are all too inclined to believe that it is beneath the dignity of the human spirit to apply oneself diligently and continuously to specific and concrete experiments and objects, and that our mind forfeits its dignity when it descends to the study, let alone the practice, of the mechanical arts. The mind here stoops to questions in which research is laborious, reflection inglorious, and exposition difficult. Such questions are dishonorable to deal with, countless in number, and of scarcely any value. (*Minui majestatem mentis humanae, si in experimentis et rebus particularibus, etc.*, from Bacon,

---

12 This entire paragraph is taken from the entry, "Art," in the French *Encyclopedia*. Beckmann occasionally omits one or more sentences. The full excerpt of the entry is quoted here and Beckmann's omissions are signaled by brackets.

Novum Organum).[13] This prejudice has tended to fill the cities with useless spectators and with proud men engaged in idle speculation, and the countryside with petty tyrants who are ignorant, lazy, and disdainful. [Such was not the thinking of Bacon, one of the foremost geniuses of England, nor of Colbert, one of the greatest ministers of France, nor, in a word, of the right-thinking and sensible men of all times. Bacon considered the history of the mechanical arts the most important branch of true philosophy. Therefore, he certainly did not scorn its practice. Colbert found the industry of the people and the founding of manufactures to be the most reliable resource of a kingdom. In the opinion of those who today can discern true worth, the state benefited no less from a man who filled France with engravers, painters, sculptors, and artists of all types, who wrested from the English the secret of the machine for producing hosiery, from the Genoese their velvet, from the Venetians their mirrors, than it benefited from those who vanquished the enemies of France and took their fortresses. In the eyes of a philosopher, a sovereign may deserve more praise if he has encouraged men like Le Brun, Le Sueur, and Audran, if he has had the battles of Alexander painted and engraved, and the victories of our generals represented in tapestry, than he would for having gained those victories.] Place on one side of the scales the actual advantages of the most sublime sciences and the most honored arts, and on the other side the advantages of the mechanical arts, and you will find that esteem has not been accorded to the one and to the other in just proportion to the benefits they bring. You will discover that far more praise has been heaped on those men who spend their time making us believe that we are happy, than on those who actually bring us happiness. How strangely we judge! We expect everyone to pass his time in a useful manner, and we disdain useful men.[14]

---

13 The quote from Bacon's *New Organon* is drawn from aphorism 83 of Book 1: "The problem has been wonderfully compounded by a certain opinion or judgement, which is deeply ingrained but arrogant and harmful, namely that the majesty of the human mind is diminished if it is too long and too deeply involved with experiments, and with particular things which are subject to the senses and bounded by matter: especially as such things tend to be laborious to investigate, ignoble to think about, crude to speak of, illiberal in practice, infinite in number and short on subtlety. And so it has come at last to this, that the true road is not only deserted but also barred and closed off; experience is badly managed, or despised and even abandoned" (Bacon, *New Organon*, 68).

14 Denis Diderot (attributed), "Art," in *The Encyclopedia of Diderot & d'Alembert Collaborative Translation Project*. http://hdl.handle.net/2027/spo.did2222.0000.139 (accessed August 16, 2023).

§9 *Materials* are what one calls the raw or already partially processed natural products which the handcrafts refine. Secondary materials are what one calls those which serve as auxiliary resources during refinement. *Hand tools* or *tools* are the various pieces of equipment with which the individual works are carried out. *Workshop* is what one calls the place which is set up for certain kinds of processing. *Technical terms* are designations and figures of speech, which are unique to individual or multiple handcrafts for their works and tools. *Wares* are the various products of the handcrafts, which one tends to call *art pieces* in the case of the fine arts.

1. The skillfulness of the handworker, and the elaborateness of the tools, mostly stand in indirect relation. The more elaborate the tools, the simpler the work.
2. The technical terms belong to the still unacknowledged wealth of our language. Most of them are very old, many are mangled foreigners, and the derivation and correct spelling of few of them have been investigated and determined. A pre-emptive apology for a few mistakes, which I have not been able to avoid.
3. It is unpleasant that some tools and works, in various handcrafts, have entirely different designations. If one wanted to work philosophically or systematically on technical terminology, one would be required to do away with synonyms rather than to introduce new names. But given that someone would have to bring this work to fruition, which, as useful as it would be, will still not take place anytime soon, then one would still have to acquaint oneself with the common language of the handcraftsmen when one wishes to give them advice, directions, and laws, and to have reports and comments from them, or when one wants to be understood by them and they by us, just as one must know well the provincial names of the plants if one wishes to make botany useful in general.
4. With all the wealth of our language when it comes to technical terms, it is nonetheless lacking an individual name for various handcrafts, objectively considered, even if it has a name for the master who pursues it, which one then tends to use, rather than the missing one. The art to dye is called dyeing, the art to paint is called painting, etc., but what should be called the art of making gunpowder, wine, starch, or gut strings? It is a difficult undertaking, to make new names according to analogy, and thereby to complete what is missing. One easily observes that in the case of such arts whose work can be expressed in German through a single verb, the name for the artist can be formed through the final syllable *er*, and the name of the art through the final syllable *ey*, for example: bake (*backen*), baker (*Bäcker*), bakery (*Bäckerey*), boil (*sieden*), boiler (*Sieder*), boilery

(*Siederey*). Many words emerge the same way, through composition, for example: beer brewery, book printery, and the like. Often the verb *make* is applied to it, for example paper-maker, stocking-maker, and this composition can often be applied, if only something contemptible did not stick to the words paper-makery, hose-makery and the like. At times, the name of the artist is derived from his ware, for example needle, needler (*Nadler*), bowyer (*Bogener*), gunmaker (*Büchsenschäfter*) and the like, but would one tolerate it if these words were given the final syllable 'ey'? In many cases one can attach the word *art* (*Kunst*), whether to the verb or to the noun, for example: varnish-art (*Lackirkunst*), apothecary-art, saddle-art (*Sattlerkunst*) and the like, but this composition will not work every time. – Perhaps this observation seems trivial to many people – nor do I want to pass it off as anything more – but Cicero did not shame himself to make the same one when he wanted to deal with philosophy and lacked the words which he needed.[15] Should technology ever be handled with the same enthusiasm in Germany, then our language would gradually accept a number of new words, just as has been adopted into philosophy, the study of nature, and mathematics.

§10 In the choice of location for a fabrick or manufacture, one has primarily to take care that the primary and secondary materials are to be had in sufficient quantity and at cheap prices, that the employee's wages are inexpensive, and that the transportation of the materials and the transfer of the wares can occur without great cost.

1. Whoever knows a fabrick or manufacture completely and thoroughly will easily be able to investigate and judge those three matters, so I will not touch upon them again in detail within the following pages.

§11 The price of the wares comes from:

1. the price of the materials
2. the interest of this capital, calculated from the purchase of the materials through the selling of the wares
3. the capital which resides in the tools and various handcraft's equipment
4. the interest of this capital, which one has to calculate as higher than usual interest because of the wear of the tools
5. the cost of the workers
6. the interest of this capital

15 Cicero is known for inventing technical terms in Latin that were translations from Greek philosophy.

7. the interest of all handwork facilities
8. the expenditure accrued during the purchase of the materials, selling of the wares, managing of the calculations, and correspondence, etc

§12 Technology is the science (*Wissenschaft*), which teaches the refinement of natural products, or the knowledge of handcrafts. Instead of just being shown in the workshops how one should follow the guidelines and customs of the master for the fabrication of wares, technology gives, in systematic order, thorough instruction as to how one finds the means from true principles and reliable experiences for precisely this goal, and how one should explain and use them for the refinement of the phenomena at hand.[16]

1. It is evident that every art has its speculative and its practical aspect: the former consists in knowing the principles of an art, without their being applied, the latter in their habitual and unthinking application. It is difficult if not impossible to go far in the practice of an art without speculation, and, conversely, to have a thorough knowledge of the speculative aspects of an art without being versed in its practice. In every art there are many particulars concerning its material, its instruments, and its application which can only be learned through practice. It is the function of practice to present difficulties and phenomena, while speculation must explain the phenomena and solve the difficulties. Consequently, only an artist who can think logically can talk well about his art. *Diderot*.[17]

3.[18] I have dared to use *technology* instead of the designation history of art, customary for some time, which is at least as incorrect as the designation

16 Torsten Meyer emphasizes the significance of this paragraph: "For the theoretical penetration of economic production and its material technics [*Stoff-Technik*] one does not need mathematics, physics, or chemistry – just technology. [Technology] defined a new scientific subject area, and claimed its equal autonomy." Torsten Meyer, *Natur, Technik und Wirtschaftswachstum im 18. Jahrhundert. Risikoperzeptionen und Sicherheitsversprechen.* (Waxmann, 1999), 148.

17 Directly lifted from the entry "Art" in the *Encyclopédie*. Denis Diderot (attributed), "Art," in *The Encyclopedia of Diderot & d'Alembert Collaborative Translation Project*. http://hdl.handle.net/2027/spo.did2222.0000.139 (accessed August 16, 2023). The idea described here is close to what Richard Sennett, in a contemporary context, refers to as "tacit knowledge": "When we speak of doing something "instinctively," we are often referring to behavior we have so routinized that we don't have to think about it. In learning a skill, we develop a complicated repertoire of such procedures. In the higher stages of skill, there is a constant interplay between tacit knowledge and self-conscious awareness, the tacit knowledge serving as an anchor, the explicit awareness serving as critique and corrective. Craft quality emerges from this higher stage, in judgments made on tacit habits and suppositions." Richard Sennett, *The Craftsman* (Yale University Press, 2008), 50.

18 The numbering skips from one to three in the first edition.

natural history is for the study of nature. The term history of the art may be what one calls the story of the invention, the progress, and the further fate of an art or handcraft, but much more so it is technology which neatly and clearly explains the results and principles of all works. At least these words are old: (*technologia*), (*technologon*), (*technologos*),[19] but certainly the Greeks were not always thinking about handcrafts, just as little as they thought with (Gk. *oikonomia, politian*) and hundreds of other words what we understand today.

4. A principal point of technology is the correct determination of primary and secondary materials which I, if I were to treat them individually, would call *materia technologica* or the study of materials. It is still scarcely considered, and still full of holes, which however need to be filled in just as well as they are in the *materia medica*, since the naturalists have taken it up. In the charter (*Freyheitsbrief*), which the mines in the principality of Halberstadt and the earldom of Reinstein received on December 23 in the year 1704, indigo still numbers among the minerals upon which it was permitted to mine.[20]

§13 One has divided the handcrafts in many different ways, for example, according to materials: into stoneworkers, metalworkers, etc., or according to the use of wares: into those for nourishment, clothing, decoration etc., or according to the various organization of the guilds: into guilded or free, etc. (§7). Still, all these divisions do not serve as the basis of technology. After many attempts it seems to me to be most advantageous to bring into one division the handcrafts whose greatest works have an equivalence or similarity in the process itself and in the principles whereupon they rest, such that the simple ones are named first and the more elaborate are named next. In such a way, much time and effort will be saved in their presentation.[21]

---

19 I have transliterated the Greek terms here.

20 Since indigo is extracted from a plant and is not a mineral, this statement needs clarification. In the entry on "Indig" in Adelung's dictionary, one can read that the Latin term *Indus* also refers to *Lasurstein* (lapis lazuli). Adelung's entry also refers to the very same charter that Beckmann does and suggests that even though indigo is included as a mineral, cobalt was probably meant, which yields a similar color. *Grammatisch-Kritisches Wörterbuch der Hochdeutschen Mundart*, Ausgabe letzter Hand (Final edition), Leipzig 1793–1801, digitalized version in the dictionary network of the Trier Center for Digital Humanities, Version 01/23, https://www.woerterbuchnetz.de/Adelung (accessed August 18, 2023), s. v. "Indig."

21 Wolfhard Weber has interpreted this paragraph in the context of Beckmann's familiarity with the work of the naturalist Johann Blumenbach: "methodologically, Beckmann emphasizes, just as Blumenbach does, the value of comparing similar things and like

1. I admit that this division also has its difficulties, but it certainly has fewer than all the ones known of before. The greatest one seems to lie in the fact that some handcrafts join together works that have very different foundations. One might think of apothecary art, bronze foundry, and organ-making. But one should treat such arts only once one has first put forth the works which pursue individually those which the former join. In lands and towns where such very elaborate handworks flourish, they have themselves already separated into many simple ones. For example, in Nuremberg the bronze founders partition themselves into formers, brass pourers, lathe operators, lantern makers, pulley makers, spigot makers, ring makers, etc. I see no disadvantage if in the same way, in technology, just as the presentation demanded, one separated a few into several more, and combined many into one. Occasionally, handcrafts have separated themselves for insufficient reasons, due to pedantry which can have no influence in technology, regardless of the fact that it has at times caused mischief and disorder, which governance often cannot resolve. To this belongs the division of the paper makers into smoothers and stampers; the needle makers and the hook makers, since the former work towards the right and the latter to the left; and the polishers into grinders and sword polishers, since the former let the grinding stone turn towards them and the latter away from them.
2. The following division, which I have developed according to these principles, is neither complete nor adequate, nor otherwise without mistake. But if one sees it as an attempt that is worth improvement and supplementation, then the effort I put into it will have been rewarded. Since within every group I place the handcrafts which have simpler works first and have tried to name afterward those which combine several, I have thus only paid attention to the handwork and not the quantity, size, and infrequency of the natural talent and knowledge of whoever wishes to be great in his art. Thus the more honored arts occasionally appear here before the less honored ones, but I do not delude myself to have drafted a hierarchy. A superficial reminder for those who want to understand me, but not a satisfying one for those who do not want to regard the arts from a single perspective, with me. Linné said: I divide the animals according

Buffon and Blumenbach, Beckmann accepts that one can conclude similarity in the *cause* of an organizational form based on the similarity of particular traits" (Weber, "Die 'Anleitung zur Technologie,'" 10). At the same time, Weber notes that this idea would only become fully realized for Beckmann in his work on a general technology published some decades later.

to the constitution of their bodies, and therefore give man a place among the mammals. Will those contradict him, who mockingly reproach him for prioritizing human intellectual forces compared to the so-called instinct of the other animals, which he knows better than many of his opponents?

§14 Natural orders of the handcrafts and arts

I.

1. Handcraft of the slaughterers, butchers[22]
2. Culinary art

II.

3. Preparation of cheese, butter
4. Oil-pressing. Olive oil, rapeseed oil, etc.
5. Train-oil boiling[23]
6. Preparation of whale sperm
7. Preparation of isinglass,[24] sturgeon glue (*Hausenleim*)
8. Glue-cooking
9. Soap-making
10. Chandlering

III.

11. Preparation of noodles
12. Wafer baking (*Oblatenbäckerey*)
13. Baking arts, black bread baker (*Grobbäcker*), white bread baker (*Weißbäcker*), fine bread baker (*Losbäcker*), brown bread baker (*Fastbäcker*)
14. Ginger bread baking (*Honigkuchenbäckerey*, *Pfefferkuchen*, *Lebkuchen*)[25]
15. Preparation of chocolate
16. Confectionary, cake baking

---

22 Beckmann lists *Schlächter* (slaughterers), followed by *Fleischer*, *Metzger*, and *Knochenhauer*, which are synonymous and indicate a butcher.

23 Train oil was derived from sea mammals, such as whales, and was used for soap-making and lamp oil.

24 Isinglass is a kind of gelatin made from fish.

25 Ebers' dictionary translates each of these as "ginger bread." Ebers, vol. 2 (1798), s.v.v. "Honigkuchen," "Pfefferkuchen," and "Lebkuchen."

IV.

17. Preparation of wines.
18. Beer brewing.
19. Vinegar brewing. Wine- and beer-vinegar.
20. Brandy brewing, Rubbing alcohol (*Franzbrantwein*) and grain alcohol.
21. Preparation of liqueur, oils
22. Aqua fortis distillery
23. Vitriol acid distillery
24. Apothecary art.

V.

25. Borax refinery
26. Potash boilery
27. Saltpeter boilery
28. Cooking salt boilery
29. Alum boilery
30. Vitriol boilery
31. Sugar boilery
32. Sugar refinery
33. Preparation of various neutral salts

VI.

34. Preparation of starches
35. – of woad
36. – of indigo
37. – of litmus
38. – of ink
39. – of carmine
40. – of painter's lacke
41. – of pastel colors
42. – of lead pencils
43. – of ultramarine
44. – of Berlin blue
45. – of ink
46. – of quicksilver precipitates

VII.

47. Tobacco fabrication. Smoking and snuff tobacco.
48. Perfume arts

VIII.

49. Wool dyeing.
50. Linen dyeing.
51. Silk dyeing.
52. Feather dyeing.
53. Fur dyeing.
54. Art of dyeing wood, animal horns, bones

IX.

55. Preparation of wall papers, flocked paper, velouté.
56. Preparation of oil cloths
57. Canvas printing
58. Cotton printing
59. Flannel printing, two faced printed flannel.
60. Copperplate printing
61. Playing card printing
62. Book printing

X.

63. Handcraft of the whitewasher, mudwaller
64. Fresco painting
65. Carton painting
66. Fan painting
67. Decorative painting
68. Art of enameling
69. Art of gold-plating
70. Preparation of Turkish paper
71. Art of illumination
72. Art of writing
73. Portrait- and landscape-painting
74. Miniature painting

XI.

75. Linen bleaching
76. Wax bleaching
77. Hair bleaching

XII.
78. Tannery
79. Tawery[26]
80. Chamois tannery
81. Art of leather dressing
82. Parchment making
83. Preparation of gut strings

XIII.
84. Preparation of leather carpets
85. Preparation of leather cannisters
86. Preparation of bellows
87. Cobbler's art
88. Art of leather strap making
89. Art of making shoelaces, art of making laces[27]
90. Glove maker, purse maker, bag maker
91. Art of saddle making
92. Art of preparing furs

XIV.
93. Preparation of sails
94. Handcraft of hat-trimming.
95. Bonnet-maker
96. Tailor's art
97. Wigmaker's art
98. Feather decorator, feather maker

XV.
99. Brushmaker
100. Pencilmaker
101. Horsehair braider

XVI.
102. Weaving, twining
103. Art of ropemaking
104. Art of making shoelaces

26 Tawing was a method for turning hide into leather.
27 In this case, leather laces for fastening clothing.

XVII.
105. Needlework
106. Embroidery with gold or silver
107. Pearl embroidery

XVIII.
108. Preparation of nets.
109. Decorative nets for clothing
110. Art of knitting, knotting
111. The weaving of stockings
112. Art of lacework

XIX.
113. Hairsieve weaving
114. Linen weaving
115. Ticking weaving
116. Twill weaving
117. Damask linen weaving
118. Lawn weaving, cambray weaving

XX.
119. Cloth weaving
120. Ribbon weaving
121. Tapestry weaving

XXI.
122. Calico weaving
123. Upholstery, fustian, canvas weaving
124. Muslin weaving

XXII.
125. Silk weaving, half-silk cloth
126. Smooth, entirely silk cloth: taffeta, ribbed silk
127. Figured cloth, silk paper
128. Patterned cloth, damask
129. Stitched fabric: fancy tweel, cloths, gold woven cloth, silver woven cloth
130. Velvet, plush, woolen plush, Manchester, Velveret
131. Silk lacemaker

XXIII.
132. The art of fulling
133. Feltmaker.
134. Art of wad making
135. Art of hat making

XXIV.
136. Art of paper making
137. Preparations of things from paste
138. Maker of puppets, dolls, masks
139. Preparation of watch cases, sheaths, scabbard maker
140. Preparation of things from paper maché
141. Fan makers
142. Book binder

XXV.
143. Preparation of pencils
144. – of sealing wax
145. – of plaster figures
146. – of pastes in sulfur and the like
147. – of images of isinglass
148. – of artificial flowers
149. – of inauthentic pearls
150. Art of wax molding

XXVI.
151. Preparation of colored pencils. Rasping.
152. Sawmilling
153. Shingle maker
154. Boxmaker
155. Basketmaker
156. Wicker chair maker
157. Matt weaver
158. Straw hat maker
159. Manufacture of fly swatters, brooms
160. Whip handle maker
161. Barrelmaker

XXVII.

162. Last-maker,[28] shingles, heels, wooden shoes
163. Cork cutter
164. Crossbow maker
165. Cartwright
166. Producer of wooden shafts for weapons
167. Wooden button maker
168. Nuremberg wares, toys
169. Carpenter
170. Turner
171. Wooden form cutter
172. Wood sculptor, image carver

XXVIII.

173. Pipe maker, pump maker
174. Pipe maker (for smoking)
175. Ring maker
176. Decoy whistle maker
177. Lathe art in wood, horn, bone
178. Buttons of coal, jet
179. Amber lathe worker
180. Rosary maker
181. Pearl drilling
182. Lathing with serpentinite, soapstone
183. Metal lathe worker
184. Grinder, polisher

XXIX.

185. Fiddle maker
186. Piano maker
187. Organ maker

XXX.

188. Whalebone processer
189. Horn-comb maker
190. Comb maker
191. Tortoise shell processing

---

28 Someone who cuts forms for shoes.

XXXI.
192. Plaster layer
193. Oven maker
194. Roof maker
195. Mason
196. Stucco work
197. Mosaics (1)

XXXII.
198. Glassmaker, windowmaker
199. Preparation of chalkboards, knife sharpeners, obsidian
200. Stone- and glass-grinding
201. Diamond grinding
202. Mosaics (2)

XXXIII.
203. Art of stamp making
204. Art of engraving and silver-plating
205. Art of stonecutting
206. Art of copper plating. Etching, black engraving

XXXIV.
207. Art of carpentry
208. Construction. Mills, bridges, etc.
209. Military construction
210. Plumbing construction
211. Construction of ships
212. Mining

XXXV.
213. Bark grinder
214. Madder grinder
215. Stone mill for masonry
216. Millet grinder
217. Spice grinder
218. Flour grinder

XXXVI.
219. Preparation of gunpowder
220. Fireworks

XXXVII.
221. Brickmaking
222. Crucible
223. Pottery
224. Pipe makers
225. Tin-glazed earthenware
226. Porcelain

XXXVIII.
227. Working with blue
228. Glassworks with green, white crystal
229. Glassworks for mirror
230. Mirror maker
231. Glass-working with quartz
232. Rosary making 2.
233. Glass painting
234. Enamel art

XXXIX.
235. Meudon whiting (crushed chalk)
236. Nuremberg sand (for ink-drying)
237. Sand clock maker

XL.
238. Coal burning
239. Turf burning
240. Ash burning
241. Bone ash burning

XLI.
242. Ochre burning
243. Zinc burning
244. Calk and gypsum burning
245. Preparation of red ashes
246. Preparation of Neapolitan yellow

XVII.
247. Preparation of lead calk
248. Preparation of green rust

XLIII.

249. Resin preparation
250. Preparation of distilled sulfur
251. Salt preparation
252. Camphor making
253. Poison making
254. Mercury sublimation
255. Preparation of cinnabar
256. Preparation of gold pigment

XVIV.

257. Ironworks
258. Anchor works
259. Anvil smith
260. Chain smith
261. Horseshoe smith
262. Drill smith.
263. Wire smith
264. Nail smith
265. Hole puncher smith (for leatherworking)
266. Saw smith
267. File smith
268. Spur smith
269. Container smith
270. Weapon smith
271. Sword smith
272. Knife smith
273. Scale maker
274. Locksmith
275. Instrument maker
276. Maker of small and large clocks

XLV.

277. Tin smith
278. Brass hammerer
279. Boiler maker
280. Copper smith
281. Tin plate maker
282. Trumpet maker

283. Brass button maker
284. Thimble maker
285. Book-cover maker
286. Bell maker
287. Gold and silver smith

XLVI.

288. Brass figure maker
289. Gold thread maker
290. Folio maker
291. Goldleaf maker

XLVII.

292. Penny maker
293. Art of making coins

XLVIII.

294. Jewelry maker
295. Jeweler

XLIV.

296. Tin worker
297. Gold plating worker
298. Preparation of gold and silver paper
299. Preparation of gold plate

L.

300. Lead worker for glassware
301. Iron wire maker
302. Brass wire maker
303. Fine brass wire maker
304. Gold wire maker
305. Lamé maker
306. Needle maker
307. Sewing needle maker
308. Fishing hook maker
309. Hook maker
310. Wool comb maker
311. Fence maker
312. Armor maker

LI.
313. Bullet and pellet maker
314. Poured iron, pots
315. Bomb making
316. Lead pouring
317. Brassworks
318. Copper tool maker
319. Tin pouring
320. Letter-type making
321. Cymbal making
322. Bell making
323. Cannon making
324. Statue casting. Statues.

> *Nulla ars non alterius artis aut mater propinqua est.* (There is no art which is not either the mother or relative of another.)
>
> TERTULLIAN, *On Idolatry*, chapter 8

PART 2

Technological Orders

ESSAY

Two Genres, Two Approaches to Technology: a Dictionary and a System

1 Introduction

The general consensus shared by those working in the field of technology around 1800 was that the sheer number of terms, related both to technological objects themselves and to the procedures for working with them, required order. Accordingly, techniques of organization were developed, a discourse of technology whose primary aim was to weigh the merits of various strategies of producing technological order. These strategies took the form of a variety of genres, including dictionaries, teaching manuals, and systems. By the end of the eighteenth century, taxonomies of technological objects and techniques had reached a degree of sophistication akin to those taxonomies of living organisms whose job was to account for objects in the natural world. And just as those scholars who worked to improve systems of nature had cause to lament systemic weaknesses and challenges posed by various animals (the platypus comes to mind), so too did the technologists complain about the inadequacy of each particular attempt to organize technological objects, even as they remained committed to the project as a whole.

One of the attempts to create technological order is the technological dictionary. This subgenre arrived on the scene in German-language publications at the beginning of the 1780s in the form of Johann Jacobsson's *Technological Dictionary* (first published in 1781). While there had been dictionaries of technical terms (understood in a very general sense) published before in various European languages,[1] Jacobsson's was different. Its orientation towards the idea of technology in the sense established by Beckmann is clear from the first pages where one finds a preface by Beckmann, given pride of place before Jacobsson's own.

Much of the biographical information known about Johann Karl Gottfried Jacobsson (1725–1789) comes from Beckmann himself. According to the information Beckmann provides, Jacobsson was born in 1725 in Elbing (Prussia). He studied in Jena and Leipzig but was required to depart prematurely because

1 Thomas Dyche's *New General English Dictionary*, for example, mentions difficult words and technical terms in the context of the the vocabulary employed by scholars in various fields.

 | DOI:10.1163/9789004549548_008

of a duel, after which he took refuge in the army. From that point onward, his personal, scholarly interests were intertwined with the various stages of his military career. He was fortunate to make the acquaintance of Peter Nathanael Sprengel and, later, Otto Ludwig Hartwig, both of whom encouraged Jacobsson's "technological investigations," as Beckmann puts it.[2]

The first edition of the dictionary was published in four volumes.[3] It contains, as the preface announces, entries in the following categories: 1) terms relating to fabricks, manufactures, handcrafts, and arts; 2) all primary and secondary materials as well as their byproducts; 3) technical terms relating to specific works; 4) tools, instruments, machines, and buildings; and 5) the wares that arise from the processing of natural materials. Given the monumental nature of the task, Jacobsson's goal was simply to cast a broad net. One can see evidence of this already in the first entry which, much like conventional dictionaries of the time, such as Johann Adelung's, simply begins with the letter 'A':

> *A* (maker of sewing needles) is the symbol of the largest sewing needles on the paper wherein they are packed. In addition, the letters of the alphabet, thus also the letter *A*, are often used instead of numbers. Thus the first sheet of paper in a book is marked with *A*. In the bookstore, [uppercase] *A* means a *thaler*, [lowercase] *a* however refers to a penny, whereby one indicates below on the title of a raw material the price of the book. An *A* on coins commonly means the first one, or the capital city or province, where they are minted.[4]

Naturally, Jacobsson is also aware that beyond individual alphabetic letters, words also acquire different meanings in particular technological contexts. For example, a relatively common word in the German language, such as *abbinden* (with the general meaning of to loosen or untie, but also to wean or to geld), can acquire specific meanings based on technological context. In this particular case, Jacobsson has four separate entries on *abbinden*, devoted to hunters, coopers (bucket makers), weavers, and carpenters. In other cases, Jacobsson will list variants of a single word, which might be spelled or pronounced differently

2 Otto Lutdwig Hartwig is also listed as editor on the title page of Jacobsson's *Technological Dictionary*.

3 Volume 1, A–F (1781); volume 2, G–L (1782); volume 3, M–Schl (1783); volume 4, Schlm–Z (1784).

4 Adelung (final edition), s.v. "A."

based on regional usage, such as "*Fahre, Fahrt, Fuhre, Furche, niedersächsisch Fore*," which are all words for the furrow made by a plowshare.

2 Beckmann on Jacobsson

Beckmann's preface expresses a genuine admiration for Johann Jacobsson and his willingness to take on the task of collecting, alphabetizing, and defining the inordinate number of technical terms in use at that time. The token criticism he feels obliged to give to avoid accusations of partiality is mild to the extreme: Beckmann notes that only on occasion does one find explanations that are "not quite adequate" or comprehensible. At the same time, the preface is clearly a vehicle for Beckmann to advance his own agenda and revisit several of the points that he made in his preface and introduction to the *Guide to Technology* from 1777. The definition he provides of technology, as "the science which teaches the manifold preparations of the products of the earth, or the knowledge of handcrafts," remains the same, and it reappears, verbatim, in the entry on "Technologie" in the third volume of Jacobsson's dictionary. Beckmann also reiterates familiar refrains about the vast number of technological objects, how the endeavor to organize them is comparable in size only to the study of nature, and that this task is made difficult by the arbitrariness of how technical terms were created, the number of synonyms in existence, and the fact that provincial words have also proliferated. Beckmann laments the impracticality of visiting all manner of workshops, even as he complains about the incompleteness of some of the written manuals of technical terms.

From a rhetorical point of view, there are two gestures that distinguish Beckmann's preface to the dictionary from his earlier writing. One is his use of a metaphor to refer to the network of technology's "auxiliary sciences" as they connect to each other and to technology itself:

> The sciences, as distinct as they might seem to novices, approach each other to the extent that their borders are difficult to recognize and to determine, so that the person who can observe them from the greatest vantagepoint believes to see that column of Phidias, upon which no joints were visible.[5]

5 Beckmann, preface to the *Technological Dictionary*, p. 107.

Phidias (480–430 BCE) was a Greek painter, architect, and sculptor. The anecdote alluded to here is of dubious provenance, but was a familiar one in the eighteenth century, notably appearing in an essay by Friederich the Great in which he critiques the political thought outlined by Niccolò Machiavelli in *The Prince*. According to the anecdote, Phidias had competed against his student, Alkamenes, to see who was the better sculptor. Each completed a large-scale statue of Athena. Whereas Alkamenes' statue was remarkable in every fine detail, the features of Phidias's statue looked rough and imprecise by comparison. Nevertheless, Phidias won the competition, because the enormous statues were meant to be viewed from afar. Distance rendered the features of Alkamenes's statue vague and indistinct, whereas the greater perspective turned the rough features of Phidias's Athena into a harmonious whole. As Friedrich the Great summarizes it: "Phidias owed his success to the study of Optics and Proportion," the rules of which "ought likewise to be observed in Politics."[6] From the perspective of Beckmann, the allusion to Phidias's statue can be understood in a continuum with his earlier writing on technology in the *Guide*. There, Beckmann endorsed a kind of double vision, where the close study of empirical objects provoked a kind of wonder that demanded the use of one's intellect, knowledge, and judgement, just as a scholarly problem would. The reference to Phidias incorporates both of these moments of close study and theoretical perspective into a single narrative, while at the same time going a step further to reflect on the interrelationship of technology and other extant sciences.

The second notable feature of the preface is its conclusion, an extended quote from Johann Matthias Gesner (1691–1761). Gesner had also been part of the faculty at Göttingen after being appointed as Professor of Poetry and Eloquence in 1734. He was a renowned scholar of the classics who had devoted his energies to lexicographical pursuits, such as the *Novus Linguae et Eruditionis Romanae Thesaurus* (New thesaurus of Roman language and learning after Stephanus) (1747) and the *Index etymologicus latinitatis* (Etymological index of latinity) (1749). Beckmann quotes Gesner at length:

> It is clear that lexicography is governed by no other rule besides that it should gather into the common reservoir whatever is profitable and should faithfully indicate the sources from which each and every piece of information was compiled. As for the rest, *no one inquires how you*

6 Friedrich the Great, "Anti-Machiavell," in *The Works of Nicholas Machiavel* [sic] vol. 1, trans. Ellis Farenworth (Thomas Davies, 1762), 622.

> *became rich; all that matters is that you are rich.*[7] Fitting for our purposes is that famous piece of advice given to the prostitute by the old pimp, "*Strip, flay, and fleece whomever you chance upon*," the reason being that '*There is no one who comes to you who is not prepared to satisfy his need from you again and again.*'[8] For dictionaries are among the most useful sort of books which everyone makes use of, with ease and convenience, for every imaginable whim, and from which they benefit not just daily, but at practically each and every hour; the sort of book that no one rashly admits what he has learned from it, and which few others, oppositely, single out for blame.[9]

The quote is taken from the twelfth paragraph of the preface to Gesner's thesaurus. That Beckmann chose it is remarkable for several reasons, not least of which is the bawdy tone (due to the somewhat unexpected comparison between dictionaries and prostitutes). Beyond that, there is the fact that through the extended reference, Beckmann acknowledges a literary kinship with Gesner and thereby a kind of parallelism with Gesner's own dictionary work, which was based on classical sources, and the present effort to establish a technological dictionary. And finally, there is the complexity of the reference itself, due to the embeddedness of the literary texts cited: Beckner quotes Gesner, who in turn refers to the Roman writers Terence (2nd century BCE) and Juvenal (late 1st/early 2nd century CE). The central message of utility, originally embedded in the context of human sexuality, is thereby transferred from human objects, to classical source material, to technological objects and, finally the object of the dictionary itself. The use and frequent contact with such a dictionary, not just daily but even hourly, suggests that, whether or not one "rashly admits what he has learned from it," its knowledge becomes a kind of prosthetic extension of our own.

3 Georg Friedrich Lamprecht's *Instruction Book of Technology* (1787)

After the translations of the prefaces to Jacobsson's *Technological Dictionary* comes the introduction to Georg Friedrich Lamprecht's *Instruction Book of Technology* from 1787. Compared to Beckmann, relatively little has been

7 Juvenal, *Satire*, 14.207. My thanks to Brandon Bark for the translation of the passage from Latin and for providing the literary references.

8 Adapted from Terence, *Hecyra* 65–9.

9 Reference will include the page number of my translation.

written on Lamprecht (1759–1810), whose official title at Halle was Professor of Law and Philosophy. He is referred to by a (near) contemporary as "the first who, through his draft of an encyclopedia and methodology of economical, political, and cameral sciences for use in academic lectures … sought to remedy a gap."[10] In general, more attention has been paid to his writings on political theory, which discuss the degree to which the state should police public morality, than to the writings on technology. He published the *Versuch eines vollständigen Systems der Staatslehre* (Attempt at a complete system of politics) in 1785, three years before his attempt to systematize technological objects. In an essay on the role of the state in German political thought of the eighteenth and nineteenth centuries, Diethelm Klippel describes Lamprecht's concerns as being oriented toward the government's role in suppressing immoral qualities (ranging from excessive luxury to begging and drunkenness) while at the same time furthering the intellectual and moral development of the populace.[11]

The translation of the introduction to Lamprecht's *Instruction Book of Technology* included in this volume has been undertaken to prevent his name from disappearing from the annals of technology altogether and, just as importantly, to help readers appreciate some of the features of his approach, which combines theoretical and practical elements in equal measure. To support these joint goals, the following pages provide an overview that also underscores what makes Lamprecht different from other technologists of his time.

When one compares the first paragraphs of Lamprecht's introduction alongside those from Beckmann's, one can see that Lamprecht neither builds up to a definition of technology, concept by concept, as Beckmann does, nor does he lead with a definition of it. Instead, Lamprecht's discursive point of departure is a reflection directed toward the text itself, a self-referential gesture that addresses what an introduction to technology is supposed to do. In response to his own implicit question, Lamprecht states that two things belong to an introduction to technology: "those general doctrines that designate what distinguishes this science from other sciences" and a "historical presentation" on the

10 Georg Gottfried Strelin, *Versuch einer Geschichte und Literatur der Staatswissenschaft* (Johann Jakob Palm und Ernst Enke, 1827), 49.

11 Diethelm Klippel, "Reasonable Aims of Civil Society: Concerns of the State in German Political Theory in the Eighteenth and Early Nineteenth Centuries," in *Rethinking Leviathan*, ed. John Brewer and Echkhart Hellmüth (Oxford University Press, 1999), 82–3. Eckhart Hellmüth has described Lamprecht's administrative zeal and drive to self-sacrifice in the name of the public good as a kind of "bureaucratic virtuosity" that was not atypical for late eighteenth-century Germany. Eckhart Hellmüth, "Der Staat des 18. Jahrhunderts," in *Aufklärung* vol. 9, part 1, ed. Günter Birtsch (Meiner, 1996), 21.

need for a science of technology and how it developed.[12] In accordance with the stated need for both a theoretical and historical approach to technology, Lamprecht divides the introduction into two main sections: a propaedeutics (*Vorbereitungslehre*) of technology and a "literary history." The propaedeutics contains chapters on the concept of technology, divisions of technology, the sources of technology, the auxiliary sciences of technology, the method of teaching and learning technology, and the use of technology. The second section, on the literary history of technology, contains chapters on 1) the history of technology and 2) a lengthy bibliography of relevant works, divided into subcategories, such as by genre (dictionaries, systems, instruction books), or by authorship (works by individuals as opposed to learned societies).

The intertextual dimension to Lamprecht's introduction cannot be overstated. Before he even offers a definition of technology, the second paragraph of the introduction (which is also the beginning of the "first main section," the propaedeutics) simply lists three bibliographic entries: Benedikt von Hermann's *Einführung des Studiums der Technologie* (Introduction to the study of technology) (1781), Lamprecht's own earlier work, *Das Studium der Cameralwissenschaften* (The study of the cameralist sciences) (1783), and Johann Krünitz's *Encyclopedia*, volume 33, page 488. What would these references have signaled to a contemporary reader? By focusing on the ways in which the three texts approach the question of technology, one can see that they each have something different to offer. Hermann's essay, for example, underscores the close relationship of the study of nature and the study of technology,[13] emphasizing the fact that technology teaches the transformation of nature products to art products.[14] However, it also gives voice to a certain terminological discontent: "technology is a science," Hermann writes, "whereof very many people still do not possess any adequate concept, even though the things which it contains are well enough known, at least by name."[15] The second reference, to Lamprecht's earlier text on the cameralist sciences, is a fifty-page essay which offers the basis for his thinking about technology, one that (as the name suggests) is still grounded in the framework of a theory of cameralist

12 P. 116.

13 "Immediately after the history of nature, or rather the theory of nature, follows technology, or the theory, which shows how products of nature are used, and to what they can be refined." Benedikt Franz Hermann, "Einleitung" [Introduction], in *Ueber die Einführung des Studiums der Technologie, oder über die Lehre von Handwerken, Künsten, Manufakturen und Fabriken* [Introduction to the study of technology or on the doctrine of handcrafts, arts, manufactures, ad fabricks] (Bernardische Buchhandlung, 1781), 9.

14 Hermann, "Introduction," 21.

15 Hermann, "Introduction," 18.

studies. The *Instruction Book* does not simply repeat Lamprecht's established position on technology, however. It offers both a more comprehensive definition and, in general, a systematic framework for organizing the various aspects of technology's theory, history, and literature. And the final reference in the list, which directs readers to the 488th page of volume 33 of Krünitz's encyclopedia, contains a definition of technology as a science that "teaches how the raw products of nature are processed and refined through human art for the requirements of life," along with the comment that only the mechanical arts belong to technology, and not the "fine arts" since they are "not interesting to cameralists, as cameralists."[16] Lamprecht's *Instruction Book* from 1787 expands the definition of technology in this case as well and also deviates both from Krünitz and from Lamprecht's own earlier position through the decision to include works in the fine arts under the auspices of technological objects.

Each chapter of Lamprecht's introduction has a certain task to perform where the concept of technology is concerned. The first chapter establishes Lamprecht's definition of technology and reflects upon different ways in which it either does or does not make sense to think systematically about technological objects. Lamprecht defines technology as:

> that science, which teaches the principles (*Grundsätze*) and means (*Mittel*) according to which and through which all those products of the earth (*Naturalien*) that, in the condition in which nature gave them to us of either no use or of only limited use, are to be processed in the best way as required for the satisfaction of various human needs, with the exception of those procedures which only afford a medicinal use.[17]

As far as definitions go, it is more nuanced than what one finds in the twelfth paragraph of Beckmann's introduction, where technology is simply defined as the science "which teaches the refinement of natural products, or the knowledge of the handcrafts."[18] In a peculiar way, Lamprecht's definition is both more general, with its inclusion of "principles" and "means," and more focused on the instrumental value of technology for human needs. At the same time, there is a certain emphasis on the corporeality, or materiality, of technological objects, as Lamprecht emphasizes in the following paragraph, where he writes that

16 Johann Georg Krünitz, *Oekonomisch-technologische Encyklopädie* [Economical-technological encyclopedia], vol. 33 (Joachim Pauli, 1985), 488.

17 P. 117.

18 P. 75.

"its objects are consequently, in all of its individual parts, products of the earth, by which all those corporeal things produced by nature are understood."[19]

What follows is a series of thought experiments in which Lamprecht tests out various ways of organizing technological objects – only to find each of them lacking. These include:

1. An attempt to organize according to "producing or fundamental" vs. other arts: By "producing or fundamental" arts, Lamprecht refers to situations where nature is "caused to bring forth productions of high quality." These productions, however, do not belong to technology, according to Lamprecht, presumably because there are no techniques of processing involved.
2. "Non-producing" arts are further divided into two categories: those which produce a lasting object, and those which do not. The former are known as handcrafts (*Handwerke*), the latter are completely excluded from technology, emphasizing Lamprecht's material bias where technological objects are concerned. As the initial definition of technology states, only handcrafts used for non-medicinal purposes belong to technological objects, in Lamprecht's opinion.
3. When Lamprecht poses the question, in paragraph 12, of whether fine arts should be included among technological objects, he answers in the affirmative. Categories of beauty and taste, for example, apply to a number of arts, and the science of aesthetics can also serve as one of the "auxiliary sciences" of technology. This decision also connects to Lamprecht's willingness, in paragraph 15, to entertain both a narrow definition of "art" (as is more common today) as well as the broader definition of art as handcraft.
4. Lamprecht rejects the distinction between free and mechanical arts since it relies on the guild status of particular arts and such status varies regionally.
5. A distinction between arts that are practiced as handcrafts on the one hand or within fabricks or manufactures on the other also does not make sense, for Lamprecht, given that the same handcraft might be practiced on a smaller or larger scale, depending on available resources.

The fine-tuned distinctions regarding what does or does not count as a technological object, which Lamprecht entertains in the first chapter, feed into more general, and equally inconclusive, musings in the second chapter on the "divisions of technology." Here, Lamprecht introduces a distinction between

19 P. 101.

a "general technology" and a "specific technology" that Beckmann will subsequently take up as well, although in greatly altered form. Lamprecht's thought that a "general technology" which would treat "the theories of all kinds of handworks" and contrast against a "specific technology" (that is, oddly enough, defined regionally, as a "European" or "German technology," for example) is difficult to defend. He also rejects, however, a distinction between "theoretical" and "practical" technology. After listing a few other, equally futile, endeavors (such as the distinction of handcrafts based on organic or inorganic materials, which quickly proves infeasible), Lamprecht's only recourse is to devise a strategy that has more to do with the concept of numerical frequency than categories of inclusion and exclusion:

> Error-free divisions as well as complete and logically precise concepts of the divided *parts* of technology seem, after all the attempts, to be impossibilities due to the innumerable quantity of completely different objects – and yet, the order in the technological system also seems to me to be the most fitting, and for the ease of study the most appropriate, if one first of all, in a *general* section, puts forth such doctrines which apply either to all, or at least to most handcrafts, and then, in the *specific* section, organizes those handcrafts according to the forces primarily operating within them, and allows the more complicated handcrafts to follow the simpler ones throughout. (paragraph 23)

Even if, by his own admission, Lamprecht is only able to come up with a kind of statistically-informed compromise, from the perspective of a history of speculative thinking about technology as a science and technological objects as a whole, his *Instruction Book* testifies to a broader, more comprehensive thinking about technology than one might otherwise be familiar with in the eighteenth-century context. The remainder of the introduction also contains glimpses of speculative technological thinking *in nuce*, such as an interest in defining the "truths of technology" (paragraph 25), the question of how to integrate "useful propositions" from auxiliary sciences (paragraph 33), the problem of how to use the science of technology for the advancement of the "general good" (paragraph 39), and the role of the technologists to reveal "the rules and principles" upon which those inventions rest which were discovered by chance (paragraph 41). With regard to the latter, Lamprecht re-introduces the analogy argument: the idea, familiar from both Bacon and Beckmann's reading of Bacon, that "the precise familiarity" with multiple handcrafts can lead to "the opportunity of making workable suggestions for applying the methods, tools, and machines of one handcraft towards the improvement of another" (paragraph 42). In Lamprecht's *Instruction Book*, then, one can see lines of continuity reaching back to prior thinking about technology and technological objects, as well as clear interest in exploring new avenues of theoretical thinking.

TRANSLATION

Johann Beckmann's Preface to the *Technological Dictionary*, 1st Edition (1781)

The present dictionary is primarily dedicated to *technology*, that is, to that science which teaches the manifold preparations of the products of the earth (*Naturalien*), or the knowledge of handcrafts (*Handwerke*). One finds in it firstly a brief, general summary of all fabricks, manufactures, handcrafts, and arts, namely, their various designations; secondly, one finds all materials that can be processed by the arts, alongside the requisite secondary materials, as well as the byproducts which occur; thirdly, one finds therein a brief yet sufficiently long explanation of all the technical terms and of the works they indicate; fourthly, one finds the tools, instruments, machines, and buildings that the artisans take advantage of, alongside the use of them; and fifthly, one also finds a reliable report of all the wares that arise from the processing of natural materials.

One can judge those dictionaries, wherein sciences are treated in individual, alphabetically ordered articles, however one wishes to, but one cannot rightly blame a technological dictionary. For it alone exhibits the usefulness which books of this kind can accomplish without causing the harm which one fears in the case of other dictionaries, perhaps not entirely without cause. This usefulness lies in the fact that those who have neither time, interest, nor talent for thorough study can easily pull together some bits of knowledge from them, whereby such people seem to themselves and to others more learned and useful than they really are. Even the few articles which can be comprehensible to readers of this genre will impart to them enough instruction that they will have no cause to regret the use of this work, but not so much that they can harm themselves and others. For the explanations of technological objects are for the most part impossible without the use of technical terms, and looking up each of these individually is no matter for those who dedicate less effort to the sciences than to whist and billiards.

Without concerning ourselves with these sorts of people, one will find the usefulness of this dictionary to be extensive and important enough. No science, with the exception of the study of nature, has more objects, and more varied ones, and for that very reason a greater number of technical terms, than technology. These terms escape even the most successful memory of those individuals who remember very well the things which are indicated by them.

 | DOI:10.1163/9789004549548_009

Additionally, the number of these words, which have for the most part been created very arbitrarily, is increased even more by synonyms and provincial words, which are not infrequently incomprehensible even to the most experienced authority. One can conceive of no more convenient aid than a dictionary wherein all these designations are collected and explained. But furthermore, no scholarly work comes close to being more difficult, laborious, and boring than the development of such a dictionary. Whoever understands this will not wonder why such a dictionary has been lacking up through this point, but rather – and even more so – wonder that a man can even be found who will dare to undertake such a project. Rather than complaining about the gaps of such a work, he would more likely be inclined to excuse them, especially when that work comes so close to perfection as the present one does.

The number of handcrafts, fabricks, and manufactures is so great that it is already difficult just to compile a complete index of them. Even greater is the diversity of their works, tools, materials, and wares. In order to understand them, many auxiliary sciences are necessary which are, in turn, of enormous extent. Without natural history (*Naturkunde*) – above all, without knowledge of minerology and botany – without a good familiarity with physics (*Naturlehre*), with chemistry, metallurgy, and many parts of mathematics and other sciences, it will in fact be impossible to get to know those objects so precisely that one could describe them correctly and comprehensibly. And even if someone is found who has acquired this preliminary knowledge, what kind of difficulties will he not encounter when he wishes to apply them to the study and explanation of technology!

There are only two sources from which he can draw. These are the workshops of the handcraftsmen and the books in which the arts are already discussed. I do not want to repeat here what I have said elsewhere about the exhausting difficulties that one encounters through the use of that first and most reliable source,[1] and I do not want to calculate the time that such a course of study demands in the school of uneducated, often unmannerly, suspicious and envious instructors, nor the patience and attention which is necessary for it. Rather, I only wish to recall the impossibility of encountering all the handcrafts – which I am taking in the most general or expansive sense – in a town, much less in a country. How can it be expected for a man to frequent this wide font of knowledge! And if he must also drink from tributaries, can

1 a) In the preface to the *Guide to Technology*. (Translator's note: Beckmann uses alphabetic ordering for his footnotes in this preface; these symbols have been retained to indicate Beckmann's own footnotes).

anyone be so uncharitable as to give him the burden of mistakes that no mortal is capable of avoiding?

The other source seems to be at first glance the most convenient one, but it is the least safe and the least abundant. If one wishes to count the number of monographs on the handcrafts that have been written down by the authors themselves, and are not compiled from other writings, one will find their number to be quite scarce and insufficient. And yet, it is extremely difficult, as I know from many years of personal experience, to bring such writings together and to use them. Many are individual essays, which have long since gone missing.[2] Many stand scattered and hidden in large and valuable works. Some are translations which, even if they are without error in terms of the description of the works, nevertheless seldom tend to have the correct German technical terms, which is the object here. In this matter, the sciences that provide instruction about works of nature and the arts have almost the same fate. Those who write about the study of nature generally find it more convenient to devote themselves to a division of natural products, and even to make up new names, rather than having to understand the best available systems and to learn the designations that have long been accepted. For the same reason, technology has from time to time new and unnecessary technical terms forced upon it by writers and translators who have not learned the correct terminology.

But even when the most industrious lexicographer has inquired at all available workshops and has excerpted from all available books, how will it be possible for him to hunt down the innumerable quantity of provincial words that no idiomatic lexicon (*Idioticon*) has yet collected! How greatly will his work be multiplied by the error-ridden pronunciation of the artisans, by their etymology and orthography, which has not yet been treated by any grammatician! For the dictionaries and linguistic theories of the living languages cover only the smallest, most general and most simple part of the languages. They are comparable to the systems of nature by the scholars of antiquity, who presented in them only the most widely known and most easily identifiable plants and animal, about which one can also certainly not say that they have been previously treated by a Linné.

So useful, but yet so monstrously large is the plan of the technological dictionary! And in spite of that, one has expanded it even more! – The terminology of all parts of agriculture, gardening, forestry, cattle breeding, hunting, and fishing, as well mining, metallurgy, trade, etc., is included here, and certainly

2 b) One should refer to my *Contributions to the History of Inventions* (Beyträge zur Geschichte der Erfindungen). Part 2, page 297.

not without reason. The sciences, as distinct as they might seem to novices, approach each other to the extent that their borders are difficult to recognize and to determine, so that the person who can observe them from the greatest vantagepoint believes to see that column of Phidias[3] upon which no joints were visible. One therefore considered it more advantageous to go beyond the usual limits of technology in order, as much as possible, to include everything that someone might expect here. In the meantime, as soon as people believed themselves to be securely over the borders, they did not go further, in order not to stray into the infinite. If therefore some technical terms from economics, metallurgy, and trade do not appear here it will not be a mistake, to the extent that these have already been abundantly collected in individual works.

When one considers the exceedingly enormous difficulties encountered in the development of such a dictionary, as they have been elaborated thus far, one will accordingly become curious to receive some information about the individual who possessed the knowledge, time, patience, courage, and the required opportunities for such a project. The following comments will therefore presumably not displease. Herr Johann Karl Gottfried Jacobsson was born in the year 1725 in the town of Elbing,[4] in Prussia. His father, a merchant, allowed him to study the foundations of the sciences in the gymnasium there and in 1743 sent him to the University of Jena in order to study jurisprudence. After two years he went to Leipzig, where he completed his academic studies. In the year 1747, he received a position with the government in Dresden, but a duel arising from a so-called matter of honor that had already taken place in Leipzig required him – because he believed his opponent to be dangerously wounded, as he in fact was – to flee. In this awkward situation, which was increased through the displeasure of his father and thus also from a lack of support, he resolved to take up service with the Saxon infantry. Soon thereafter, King August[5] transferred him to the calvary, where he rose to the position of sergeant. In 1755, he accompanied the king to the parliament in Warsaw, where a severe illness kept him from profiting from the various advantages that his position afforded him, for which reason he finally saw himself forced to take his leave, and subsequently to begin a new career in the royal Prussian military service. He became a subordinate officer with the Raminian regiment, participated in the campaigns of the penultimate war,[6] but nevertheless saw the hope

3 Phidias (*c.*480–430 BCE) was a Greek painter, architect, and sculptor.

4 The town of Elbing (1733–1763), located in Prussian territories around 1800, is currently the Polish town of Elbląg.

5 King August III of Poland was also known as Frederick Augustus II.

6 Most likely the Seven Years' War (1756–1763).

of greater fortune in military service disappear, and he therefore decided to return to the sciences, for which he had retained fondness during all of the distraction. In order to approach them more quickly, he did not hesitate to return to the military role and throw himself into the onerous service of the subordinate officer, since as musketeer he could arrange more time and quiet for studying.

Among the scholars with whom Herr Jacobsson became acquainted during that time was Herr Peter Nathanael Sprengel (a native of the Mark, at that time teacher at the *Realschule* of Berlin), by whom he was encouraged to pursue technological investigations. The superior Counsellor of Consistory (*Oberconsistorialrath*) Hecker had, namely, at the time the *Realschule* was founded,[7] made the arrangement that knowledge of handcrafts should also be taught, and this teaching appointment was granted to Herr Sprengel, who subsequently published in 1767 the first part of the *Handcrafts in Tables* (*Handwerke in Tabellen*), with help of his colleague, Herr Mölling.[8] Instead of Mölling, Herr Jacobsson helped Sprengel gather materials for the second part, during the process of which, however, superior Counsellor of Consistory Hecker died, and Herr Sprengel was called to assume the preacher's office at Gentin, near Magdeburg. At Herr Sprengel's insistence, Herr Otto Ludwig Hartwig (from the Uckermark, where his father had been preacher) assumed, as instructor at the *Realschule*, the continuation of the technological work in 1768, which he immediately turned in a more advantageous direction. Herr Jacobsson doubled his efforts thereby to learn precisely for himself in the workshops all handcrafts that should be described, in order to provide even more, and more important, contributions, which are used advantageously up through the eleventh volume. When, in 1774, Herr Hartwig assumed the position of priest in Buchholz, in the Potsdam Inspectorate, and had to work through the last three volumes of the work far from Berlin, the seat of the arts, Herr Jacobsson expanded his own reports, which he had already included, and was also helpful to Hartwig in the revision of the first and second parts.

In the course of this endeavor, Herr Jacobsson undertook a more complete description of all textile manufacture,[9] which, immediately after the publication of the first part in 1773, received the praise that it still received in 1775,

7 The *Realschule* was founded in 1747. A *Realschule* is a secondary school which focuses on practical studies.

8 This multivolume work, first published in 1767, comprises an important part of the prehistory of the discourse on technology in Germany. It lists the materials, specific terminology, and techniques pertinent to the most widely practiced arts and handcrafts.

9 c) *Schauplatz der Zeugmanufakturen in Deutschland.* 4 Bände, gr. 8. (8 = octavo).

once the publication ended with the fourth volume. The bookseller Kanter in Königsberg was thereby encouraged to task Jacobsson with the translation of the excellent work, *The Art of Carpentry* (*L'art du menuisier*) by M. *Roubo*[10] for the *Theater of the Arts* (*Schauplatz der Künste*),[11] which he also, disregarding the great difficulties which such a labor is subject to, has successfully completed, even though it is not yet printed. Herr Jacobsson has also lent his support to the exemplary *Description of the Residence Cities Berlin and Potsdam* (*Beschreibung der Residenzstädte Berlin und Potsdam*) by Herr Nicolai.

In the course of these occupations for the public good, Jacobsson arrived at the resolution to develop a complete dictionary of all technical terms used in fabricks and manufactures, having had occasion enough to lament the lack of one. He openly promised such a work in the last part of his *Theater of Textile Manufactures in Germany* (*Schauplatz der Zeugmanufakturen in Deutschland*), whereupon I could not hold back my pleasure, announcing it in the *Physical-Economical Library*, volume 7, page 575, because I was at that time already convinced that apart from him few people could be in the position of accomplishing this great service to the more useful sciences. In the meantime, this plan was interrupted in 1778 by the last war. Herr Jacobsson once again participated in the campaign, but he acquired the desired leave in the winter quarters and returned at once to Berlin and the work he had already begun.

For this work, it was an uncommonly advantageous circumstance that precisely that book shop, for which Germany has to thank for the first and, as I believe, still very erudite general dictionary of its language,[12] also took over the publication of the technological dictionary. Herr Nicolai helped the author improve the great project and stake out its limits, showed him the most useful writings, acquired them and procured contributions for him,[13] gave him support, and alleviated for him the arduous undertaking in various ways, particularly by making him acquainted with and providing him with the necessary books. Herr Nicolai prevailed upon Herr Hartwig once again to join his knowledge and industriousness with the works of Herr Jacobsson for the common good and, according to the latter's own wish, to be afforded the honor of

10 André Jacob Roubo (1739–1791) was a French carpenter and cabinetmaker.

11 The reference is to a series of books published under the title *Schäuplatz der Künste und Handwerke*. A total of 21 volumes were published between 1762 and 1805. Each volume contained translations of the works of French scholars on individual handcrafts.

12 d) Namely, J. L. Frisch, *Deutsch-Lateinisches Wörterbuch* [German-Latin dictionary].

13 e) For example, Herr Nicolai has procured a few copies of the *Chemnitzischen Bergmännischen Wörterbuchs* [Chemnitz miner's dictionary] interspersed with papers for a few people knowledgeable about mining and and metalworking and also asked Herr Jacobsson for the favor of improvements and additions.

looking through the manuscript and improving it as much as possible, and to take care of the edition in such a manner.

We have these men and the coming together of these events to thank for the present work, which far surpasses all others of a similar kind, not only in our language, but also, as far as I can see, also in all other languages, and which some day, for the facilitation of the most useful knowledge, can be improved and expanded without great difficulty. Because I am no hired panegyrist, I also do not hesitate to mention the imperfections that I believe to find in this first part. Though I may thereby lighten the detestable efforts of those ungrateful persons who would rather point out the faults than the merits of a work, thus will I also, among the proper experts, justify my sincere praise, which would otherwise carry no weight.

Without displaying unnecessary examples of the unavoidable incompleteness, whose improvement is anyway already anticipated, I admit that some explanations here and there are not quite adequate, are not comprehensible. – Whoever would wish to dare deliver so many thousands of definitions individually, without the aid of systematic connection, and without reference to illustrations[14] to everyone, completely comprehensibly, would truly deliver thereby the most vivid proof of his unsuitability for such a work. – I acknowledge that a few of the definitions and descriptions of the materials did not quite succeed, but I also recall that the study of materials and wares has at the present moment received little attention which certainly cannot occur at the same time. I acknowledge that the articles are not universally of the same detail and that one can notice that the manufactures of cloth and the art of weaving are better known to the author. Finally, I acknowledge that here and there superfluous articles have slipped in which, since no one will look for them, will certainly not harm, but will also not be useful. All of these small faults, however, decline in the course of the work and disappear entirely by the conclusion of the first volume. Finally, I mention once again that Herr Jacobsson hopes to publish the book in three volumes, that he cannot yet determine the length of the last volume, and that after the completion of the entire alphabet, supplements should still follow. I wish that patrons and friends of the technological science might grant their support to this endeavor. I conclude with the words

14 f) Certainly much would be gained, for the author as for the reader, if one were to be able to include a volume of illustrations. That French and English lexicographers and publishers can securely manage this is certain; they can expect more wealthy, and more numerous buyers, and even our countrymen desire their wares; but whether these buyers will be as generous towards a German product of not less worth, this is a question that I cannot answer.

that our blessed Gesner spoke with reference to his Stephanish dictionary:[15] "It is clear that lexicography is governed by no other rule except that it should gather into the common reservoir whatever is profitable and should faithfully indicate the sources from which each and every piece of information was compiled. As for the rest, *no one inquires how you became rich; all that matters is that you are rich.*[16] Fitting for our purposes is that famous piece of advice given to the prostitute by the old pimp (adapted slightly), "*Strip, flay, and fleece whomever you chance upon*," the reason being that '*There is no one who comes to you who is not prepared to satisfy his need from you again and again.*'[17] For dictionaries are among the most useful sort of books which everyone makes use of, with ease and convenience, for every imaginable whim, and from which they benefit not just daily, but at practically each and every hour – the sort of book that no one rashly admits what he has learned from it, and which few others, oppositely, single out for blame."

---

15 In Beckmann's preface, the following quote is in Latin, for which I've substituted the English translation. I am very grateful to Brandon Bark for translating the passage as well as the reference to Terence. The original text is as follows: "apparet lexicographo non aliam esse legem positam, quam ut congerat in communia horrea, quidquid est bonae frugis, fideliterque ipsos fontes, unde quidque descriptum sit, indicet. De caetero *unde habeas nemo quaerit, sed oportet habere*. Pertinet ad nos praeceptum illud veteris lenae meretrici datum, inflexumque leviter, *spolies, mutiles, laceres, quemquem nactus sis;* ratio est in promtu, *Nam nemo quisquam ad te venit, Quin ita paret sese, abs te ut suam vicissim indigentiam expleat*. Est ea fere sors librorum utilissimorum, ut illis pro sui animi lubidine, facile et comode utantur omnes, et inde iuventur non quotidie, sed singulis fere horis; nemo temere, quid inde didicerit fateatur, pauci aliter, quam in reprehensione nominent" Johann Matthias Gesner, "praefatio" in *novus linguae et eruditionis romanae thesaurus post Stephani* (new thesaurus of Roman language and learning after Stephanus) (Caspar Fritsch, 1749), unpaginated.

16 Juvenal *Satire* 14.207.

17 Adapted from Terence, *Hecyra* 65–9, "And that's why I constantly urge and exhort you never to take pity on a lover but strip, flay, and fleece everyone you get ... Not one of them, I assure you, comes to you without the intention of talking you into satisfying his desires at the lowest possible price."

TRANSLATION

Johann Jacobsson's Preface to the *Technological Dictionary*, 1st Edition (1781)

In February (*Hornung*) of the previous year I made known to the public, through a publicly printed announcement, the plan and contents of this technological dictionary, of which I here deliver the first part. With regard to its composition I refer my readers to the preface by Herr Professor Johann Beckmann of Göttingen. Since the great knowledge of this famous man, particularly in this subject, is generally known, I have from time to time sent him the printed pages of this first part and asked him to communicate his impartial judgment to me. So that this judgement would also be made known to the public, he has openly pronounced it here in the prior preface, for which I convey to him the most obliged and devoted thanks.

From this preface, my readers will learn about what is good as well as what is lacking in this work, because in it everything has been judged according to the strictest truth. There remains for me nothing else to say, only that I will give my readers an account of the inner workings and external organization of this work with few words.

First, every person will easily see that, given this immeasurable field of materials, it is no small matter for a single person to treat everything with the most precise perfection (since entire societies should rightly work on such an undertaking), and no less that it would be impossible to collect everything without exception. Thus, there is no reason to wonder that various articles are still lacking – which I myself belatedly discovered after it was too late for printing – and I will flesh them out at the end of the work in a supplement. A work of such far-reaching scope cannot be well-executed without any support, and yet I have had less than I wished for, except for being so fortunate that a man took over the printing who, due to his extensive knowledge, was in the position to determine the plan for this work more closely with me and to procure for me the printed sources from which I have had to draw when it pertained to matters that I could not see with my own eyes. However, I have not just relied on these sources, but rather my many personal experiences in this area had much to contribute to the execution of this work. Since for more than ten years I have visited the artisans, manufacturists, and handcraftsmen of the city of Berlin so rich in artisanal workers, manufactures and handcraftsmen, and have undertaken investigations into their works, I have thereby been

 | DOI:10.1163/9789004549548_010

put in the position to describe most things not simply according to what I read, but also as I have seen it.

Beyond that, Herr Pastor Hartwig of Buchholz, who is favorably known for the extensive knowledge he has gained in this field, was kind enough to not only look over my manuscript and improve it where necessary, but also to enrich this volume with many additions, for which I here openly offer him my thanks.

By the same token, I cannot neglect to offer my most sincere thanks to the assessor Herr Rosenstiel of the highly praiseworthy royal department of mining, since he provided me, with respect to metallurgy, mining, and smelting works, many additions and improvements by talented experts, whereby I was put in the position of avoiding many mistakes and inaccuracies in these fields.

I have taken care that these articles are always positioned where they need to be found according to alphabetical order. I have considered diphthongs only as simple letters and accordingly not brought them under a particular rubric, as has up to now been the case in such works.

If I, in the description of a machine or of an ingenious tool, at times need to go on extensively, then one cannot reproach me for it, since it is not possible to accomplish this in few words without drawn figures of them. For experts of such machinery this is naturally superfluous, except that I have not only written for scholars, but rather also for such people who have absolutely no idea of this, and accordingly want to learn. Thus, I naturally had to allow myself to write extensively on occasion, in order to make such things comprehensible to such an individual even without a figure. Yet, I often felt that without the required drawing I would not be as completely comprehensible as I would like. The publisher is thus not averse, if the present work is well received, to allowing a future volume with the relevant copper engravings to follow.

However, in order to alert the reader in advance as to where he can see such machines and tools in copperplate printing for all the ingenious and complicated things, I have included references to the books where they are depicted by a copper engraving which one can consult if one has the opportunity. In order to save space, I have set the titles of the writings with only the initial letters, of which I have given a particular explanation at the end of this volume. Also, in order to save space, I have not included for every noteworthy article the sources from which I have in part gathered the materials, and where one could read more extensively about it. On the other hand, I will, instead of doing so according to the order of the index of arts and handcrafts already familiar from the preceding announcement, attach at the end a systematic index of all writings I have used for this work, whereby at the same time I have noted the page numbers for every writing wherein there is something from this or that art to consult. This index is, at the same time, to be treated as a small technological library, whereby Herr Professor Beckmann has most kindly promised his support, since he himself possesses an extensive erudition in writings of this kind. To be sure, I have let it be known in my announcement that there will be only two volumes of this work. However, while working on them I quickly realized that it would not be possible to limit the project to two volumes. The vast number of materials simply did not allow it, otherwise I would have had to leave out much, whereby the work would have only become incomplete. In addition to this, I have, at the persuasion of several learned friends, taken up more to work through than I indicated in my plan: whereby primarily trade, agriculture, warfare, culinary arts are included. Yet I have, from most of these newly included areas of study, chosen the most noble ones, for if I should want to include everything from these arts, for example, the art of cooking, then a single volume would have come of that alone. Not to mention the fact that we already have complete descriptions of it.

For these reasons, everyone who wishes for something complete and precise will not be displeased to see that the entire work will certainly be three volumes strong. On the other hand, I will exert myself in all seriousness for the greatest possible brevity, so that it will not become more. I still hope to achieve this, even though the first volume only goes through F. At the end of the work will come a supplement of belatedly discovered articles. In order to complete it as soon as possible, the second volume should be finished at the end of this year, the third volume however in June of 1782, or directly after the Leipzig Easter book fair, God and health permitting. In this way no great delay will arise from the expansion of the work, and as regards the additional costs, I believe that every reasonable reader will not mind it, since the completion of the work made it impossible to limit it to two volumes, and since beyond that the cost for advance payment set by the publisher is extremely cheap. I just need to ask my readers for indulgence regarding the few mistakes which have crept in. Since I could not look over the pages myself, due to the remoteness of the printing location, but rather had to leave the proofreading to someone else, it could not be avoided that here and there small mistakes should creep in which no one other than the author himself could notice. Still, I have also tried to point out as many as possible.

I affirm that I will exert all my powers to offer as much completeness as I can, also in the following parts of this work. I hope that insightful readers will already notice that the second half of this first volume has a distinct advantage compared to the first half. I naturally would have wished that nothing incomplete would have slipped in right from the beginning. But whoever has the slightest idea of the laboriousness and of the difficulty of such a work will realize that in the beginning, before everything is established, even more difficulties are accrued. Thus a reasonable reader will excuse the fact that despite all care some things have been left out of the beginning, and that in letter 'A' a few articles appear twice. One will not find this in the following volumes.

I leave the further judgement of my work to the public. It can decide whether or not the work turns out according to my promise. I naturally wish, after I have spent such an unspeakable amount of time on it, that it should receive the applause of judicious men, which would be the sweetest reward for my bitter labor. Yet I will also gladly hear reasonable admonitions and if I find that they are justified, accept instruction.

With this opportunity, I will put in a parenthetical word, that Herr Varrentrapp and Herr Wenner have inserted a bitter announcement against it in the all the newspapers because I had commented, in my announcement, that few technological matters are to be found in their printing house's forthcoming so-called *German Encyclopedia*. They accuse me as if I wished to disparage this work, which was certainly not my intention. With the announcement of my work I had to show the public that such a work like what I wished to deliver is not yet available in our language. To this end, I briefly indicated what we have of similar works, in order to prove that mine would be of a completely different kind. For that purpose I did not, however, disparage other works. The conceived encyclopedia has, for the many sciences that it wants to include, relatively limited coverage, so that it is therefore no wonder if little appears in it of technological definitions. Meanwhile it seems that my pointing out of this deficit has not been entirely useless, since I have actually observed in the most recent volume more of these materials.

Incidentally, since I do not love useless quill wars, I will never allow myself to indulge in extensive discussions about these and similar matters.

Berlin, the 20th April, 1781.
Johann Karl Gottfried Jacobsson

TRANSLATION

Introduction to Georg Lamprecht's *Instruction Book of Technology*, 1st Edition (1787)

§1 To an introduction to technology belong, in part, those general doctrines that designate what distinguishes this science from other sciences and, in part, the historical presentation of how and why one first began to work on this science, how it gradually developed more and more and became more complete, and which writings contributed the most to this.

§2 Technology will be divided here into two sections. First: the propaedeutics of technology, and secondly: its literary history.

First Part: Propaedeutics

1) Hermann[1] on *die Einführung des Studiums der Technologie* (the Introduction of the Study of Technology). Vienna 1781.
2) v. Lamprecht[2] *Ueber das Studium der Kameralwissenschaften* (On the Study of the Cameralist Sciences). Halle 1783.
3) Krünitz, *Encyclopädie*, vol. 33, 488.[3]

§3 The propaedeutics designates that which distinguishes technology from other sciences, and collectively, the materials that belong here can be most comfortably divided into the following six chapters: 1) on the *concept* of technology, 2) on its *divisions*, 3) on its *sources*, 4) on the *method* of *instructing* and *studying* it, 5) on its *auxiliary sciences*, and 6) on its *uses*.

1 Benedikt Franz Josef von Hermann (sic) (1755–1815) was an Austrian earth scientist.

2 Georg von Lamprecht (1759–1810) was a German pedagogue and scholar of economics and technology.

3 Johann Georg Krünitz (1728–1796) was a German encyclopedist. Volume 33 (1785) of his *Economical-technological Encyclopedia*, on page 488f., contains a definition and brief description of technology discussed in the introductory essay to this translation.

 | DOI:10.1163/9789004549548_011

First Chapter: Concept of Technology

§4 Technology, history of the arts, town economy, fabrick science, is that science, which teaches the principles and means according to which and through which all those products of the earth (*Naturalien*) that, in the condition in which nature gave them to us of either no use or of only limited use, are to be processed in the best way as required for the satisfaction of various human needs, with the exception of those procedures which only afford a medicinal use.

§5 Its objects are consequently, in all of its individual parts, products of the earth, by which all those corporeal things produced by nature are understood, which are found individually under human control, and which are still in the condition in which nature made them. Even if a few of them, as nature gave them to us, can be applied towards the satisfaction of various needs, one cannot use most of them at all without first processing them, and those few that can be used cannot be applied for any secondary purpose without such processing.

§6 The processing of these products of the earth is *either* comprised of giving them another external form, *or* of removing from them a material useful to satisfy a particular human need from the other materials with which nature had bound it, *or* in that several natural products of one kind, *or* even several of various kinds are bound with one another.

§7 Under *art* in the broad sense is understood every occupation that is undertaken, according to certain precepts and rules, with a skill achieved through practice. The *theory of an art* encloses all the rules and precepts according to which such an art is to be practiced for the attainment of its purpose.

§8 Since *technology*, in that it teaches the processing of products of the earth, deals with *theories of the arts*, the vast (*ungeheuer*) number of arts which people have gradually invented is therefore to be divided into particular classes, and it is to be shown, which of those themselves belong in the area of technology.

§9 The most general division of the arts is into *producing* or *fundamental*, and *other* arts. Under the first category is understood all those through whose practice in part nature is given cause to bring forth products of the earth of the best quality, and in part mankind appropriates what has been brought forth.

The theories of these (producing arts) do not belong at all in the area of technology. But also the theories of the *not producing* arts are not all parts of this science.

§10 One can further divide them into such arts through whose practice no remaining corporeal product issues, such as, for example, the art of music, the art of fencing, and others where this is the case. For the *former* arts we have no appropriate general name. However, one can fittingly call the arts of the latter kind *handcrafts* in the most general sense. The former are completely excluded from technology. Of the latter, one distinguishes further between the *apothecary art* and *handcrafts* in the narrower understanding. The former have, as their object, the preparation of effective medicines, the latter however concern themselves with preparing, in part from entirely unworked products of nature, in part from those already processed, all other means of satisfying human needs. The theories of handcrafts alone belong in the area of that science which is called technology.

§11 Just as, on the one hand, one cannot relegate all theories of the arts to technology, so too, on the other hand, must the theories of those arts not be excluded which are, in fact, scientifically handcrafts, even if also in daily life they, for good reasons, are distinguished by more exalted names.

§12 Accordingly, the theories of the so-called *fine arts* need not be removed from technology. Insofar as through these arts natural products (*Naturerzeugnisse*) are processed, and permanent corporeal products are produced, they belong to technology. The justification that certain rules of beauty and taste are required for their practice is not sufficient to bypass them completely, in that the same justification could also be applied to many other arts which are completely reckoned among technology. However, just as one, for the complete understanding and agreeable practice of most handcrafts, also requires knowledge from other sciences, so would aesthetics, which developed the necessary rules of beauty and of taste for its practice, be an auxiliary science for technology.

§13 One further distinguishes between the *free* and the *mechanical* arts, in that the statutes of the guild or trade organization have been introduced in some arts, but not in others. However, since very disparate statutes occur not only in various lands, but also in the individual towns of a particular land, this distinction is thus much too arbitrary for one to exclude the so-called free arts from technology.

§14 The distinction that has recently been made in daily life between *handcrafts* in the most narrow sense and *manufactures* and *fabricks* comes even less into consideration. This distinction only depends upon the manner in which a handcraft is practiced. The latter two designations are always given to a handcraft when it is exercised on a large scale, so that among several workers, each one only concerns himself with a single labor out of the various workings of the raw material, and one thereby works directly into the hand of the next. One tends to call it a fabrick when work is done with fire and hammer, and a manufacture when work is done by hand or other tools and machines.

§15 Finally, it is to be noted that one tends in daily life to call those handcrafts *arts* which in practice 1) require many natural abilities and learned scientific knowledge, 2) those, which process gold, silver and jewels, and 3) various other only recently invented handcrafts, in whose case the guild constitution has not been introduced. The theories of these belong as well under the purview of a complete technology.

§16 From technology – which, given that it is now possible to express oneself according to the concepts determined in the previous paragraphs, teaches the *theories* of the *handcrafts* – the following sciences, though closely related to it, are to be differentiated. 1) The *history of the arts*, which confers an instruction of the inventions, the general improvements, and other fates of the arts in general as well as, in particular, the handcrafts.

§17 2) Handcraft law,[4] which determines the laws and obligations which, according to laws or tradition, pertain to the various handcraftsmen or which they are bound by duty to do; 3) the civil government of the handcrafts, which teaches the principles, means, and structures according to which, and through which, the regent has to govern the handcrafts for the general good of the state; 4) the science of economy (*Haushaltungsrecht*), which sets up general principles and means which one depends upon in the attainment, preservation, and spending of one's private fortune, as well as specific guidelines

4 Krünitz defines *Handwerksrecht* in terms of various legal entities that bear directly on various aspects of the practice of handcrafts, emphasizing that they must be general enough to apply to all handcrafts. The various legal entities include regional laws and ordinances, the ordinances of the particular handcrafts, imperial laws (*Reichs-Gesetze*), trade customs, as well as Roman and Papal laws. Johann Krünitz, *Oekonomisch-technologische Encyklopädie* (Economic-technological encyclopedia), vol. 21 (Joachim Pauli, 1780), 556–7.

of the same kind for all various sorts of persons pursuing a trade, as well as for handcraftsmen.

Second Chapter: Divisions of Technology

§18 One teaches a *general* technology if one treats, in the system of technology, the theories of all kinds of handcrafts in a narrower sense that exist in the world, without distinction. One teaches a *particular* technology, for example, a European or German technology, if within it merely the theories of those handcrafts which can only be practiced in a certain land are introduced.

§19 Also possible, but never advisable, is the further division of technology into the *theoretical* and *practical* part. In the former, all of those technological truths are handled which do not immediately apply to the way in which artisanal wares are processed, and it could be divided into three sub-partitions: 1) the *study of materials*, 2) the treatment of *tools* and *machines*, 3) the *study of wares*. This would teach the way in which artisanal wares are processed out of the indicated materials with the described tools and machines.

§20 The divisions of handcrafts into *free* and *mechanical*, into *actual handcrafts* and *manufactures*, and *fabricks*, *arts*, and *fine arts*, are much too arbitrary and unessential for one to base the division of the technological system upon it.

§21 Some instructors of technology have tried to introduce and order the theories of handcrafts in the technological system according to the *materials* which are processed by *handcraftsmen*, namely, according to the three *realms of nature*, 1) those who process products of the *mineral realm*, 2), the *plant realm*, 3) and the *animal realm*. However, a single handcraft often processes materials of varied kinds, and often the processing of materials of various origins by means of various handcrafts almost corresponds, so that it would cause difficulty for the study of technology if one were, in accordance with this system, to divide such similar handcrafts from one another.

§22 In addition, technology is divided by other people according to the *usage* of various wares which the handcrafts provide, and five classes have accordingly been created, 1) such handcrafts which provide for the means of satisfying the needs of human *habitation*, 2) their *clothing*, 3) their *household objects*, 4) their *sustenance*, and 5) those which provide for the other means of satisfying human needs. However, in part there is the difficulty with this division that many handcrafts provide means of satisfying more than one of the aforementioned requirements, and in part this division also does not in the least facilitate the study of technology.

§23 Error-free divisions as well as complete and logically precise concepts of the divided *parts* of technology seem, after all the attempts, to be impossibilities due to the innumerable quantity of completely different objects – and yet, the order in the technological system also seems to me to be the most fitting, and for the ease of study the most appropriate, if one first of all, in a *general* section, puts forth such doctrines which apply either to all, or at least to most handcrafts, and then, in the *specific* section, organizes those handcrafts according to the forces primarily operating within them, and allows the more complicated handcrafts to follow the simpler ones throughout.

§24 The specific part of technology would therefore contain both of these sub-partitions: 1) the theory of the *chemical*, and 2) the theory of the *mechanical* handcrafts. The former would deal with all those handcrafts in which the products of nature are processed according to the principles of chemistry (*Scheidekunst*), since the chief matter relies on one or the other operation or force of nature that is engaged, and accordingly guided, by the art. The latter, however, would teach the theory of all remaining handcrafts, in which artisanal wares are brought forth without further influence of the forces of nature, merely through machines, tools, and manual labor, according to the principles of mechanics.

Third Chapter: On the Sources of Technology

§25 The truths of technology are based in part on experiences, and in part on conclusions from them and correct physical, chemical, and mathematical principles.

§26 Thus, *firstly*, experiences are to be used, and certainly the experiences of others in part, in that one reads and uses the best technological writings, and in part one's own experiences, since one visits the workshops of artisans, collects experiences on journeys in such lands where handcrafts primarily bloom, and even conducts experiments where possible. With this is, *secondly*, one's own contemplation to be joined, and improvements are to be added from physical, chemical, and mathematical principles, and from the similarity of procedures in various arts, for example.

Fourth Chapter: On the Auxiliary Sciences of Technology

§27 Under auxiliary sciences of technology are understood those sciences which stand in such a relation to it, that without them one would neither be able to thoroughly learn technology, nor be able to apply or to practice it properly.

§28 To such auxiliary sciences belong 1) *natural history*, which teaches people to know the products of nature (*Naturprodukte*) by class according to certain external conditions, in that the most appropriate processing of these products is based on a more precise knowledge of them, and often the nature expert, instead of using the products of the earth (*Naturalien*) that have traditionally been used for a certain goal, is in the position to suggest others as more beneficial for that purpose.

Vandelli *diss. de studio historiae naturalis necessario in medicina, oeconomia, artibus et commercio* (dissertation on the study of natural history necessary in medicine, economics, the arts, and commerce) Lisbon 1768. 8.[5]

§29 2) *Physics* and *chemistry*, which instruct in the precise knowledge of the qualities, component parts, and effects of the natural bodies, are undeniably indispensible, both for the thorough study of technology, and for its appropriate application. From them, technology takes the most noble proofs and truths that it puts forth, and they teach the observations, experiences, and operations of nature, upon which the entire method for the preparation of artisanal objects is based, and without a thorough knowledge of these sciences one is in no position to contribute to the development and improvement of handcrafts.

Gmelin *Einleitung in die Chemie.* (Introduction to chemistry) Nuremberg 1780, pages 21–26.[6]

Succow *Vom Nutzen der Chemie zum Behuf des bürgerlichen Lebens und der Oekonomie.* (On the usefulness of chemistry for the advantage of civil life and the economy). London 1775.[7]

---

5 Domenico Agostino Vandelli, Italian naturalist (1735–1816). This work was published jointly with another dissertation. The full title is: *Dissertatio de arbore Draconis, seu Dracœna. Accessit dissertatio de studio Historioe Naturalis necessario in Medicina, Economia, Agricultura, Artibus et Commercio* (Galliard, 1768).

6 Johann Friedrich Gmelin (1748–1804), German naturalist. The pages in question describe the broad application of chemical concepts for the arts.

7 Georg Adolf Succow (1751–1813), German physicist.

Wallerii. diss. de nexu chemiae cum utilitate reipublicae. (dissertation on the connection of chemistry to public usefulness).[8]

§30 An auxiliary science that is important to an equal degree for the technologist is *applied mathematics*, primarily due to its parts – mechanics, hydraulics, hydrotechnics, and hydrostatics – according to whose principles alone one often can correctly understand and judge the mechanism and use of the majority of very ingenious and complicated machines which are utilized in almost all handcrafts for the preparation of artisanal wares. Without knowledge of this kind one is neither capable of recognizing mistakes in commonly used machines, nor capable of suggesting improvements.

Fifth Chapter: On the Method of Teaching and Learning Technology

§31 The general rules, which are usually to be followed when teaching and learning the sciences, do not belong here, but rather only those that particularly matter for the instruction and study of technology, in order to develop and master a thorough knowledge of this kind. Above all, however, one must pay careful attention to the technological instruction at universities for the education of future cameralists.

§32 First and foremost, it depends here 1) on technology being taught completely, insofar as instruction will be given on all the handcrafts, manufactures, and fabricks, which are either abundant sources of national profit in the state in which one wishes to educate cameralists, or can at least become advantageous sources of profit (*Nahrungszweige*). The most simple handcrafts, however, as well as those first preparations of raw products which the agriculturalist, who gains them, procures at the same time, and finally those preparations with which the chemist is only concerned to a small degree, and which would not really be advantageous to pursue on a large scale, can quite fittingly be passed by in the academic lecture on technology, even if they cannot be left out of a complete system or dictionary of this science.

§33 Since most works in the various handcrafts rely upon physical-chemical and mechanical principles, thus 2) in the presentation of technology the propositions necessarily borrowed from these sciences must be introduced at the right place, and it must always be clearly explicated, upon which

8 Johan Gottschalk Wallerius (1709–1785), Swedish chemist. The full title is *Dissertatio chemica de nexu chemiae cum utilitate reipublicae* (1751).

principles the designated method in the preparation relies, and from where the former or the latter effects emerge.

§34 In the systematic treatment of all individual handcrafts, manufactures, and fabricks, 3) these questions must be primarily addressed in spoken presentation: 1) what are the primary and secondary materials, what are their properties, and what are their prices? 2) in what way are they processed, how do the various works follow from one another, what principles do they rely upon, which tools, machines, and techniques (*Handgriffe*) does one use, and which could be used advantageously, instead of the usual ones? 3) what are the defects and good qualities of wares, and their signs? 4) for what are the completed wares used? 5) what are the technical terms (*Kunstausdrücke*) for the works, tools, machines, and defects of the wares? 6) what are the most important home-fabricks (*Landesfabriken*) of the kind, where do they have the most sales, and at which price do they deliver their wares? 7) what kind of capital, and what kind of a building, and how many workers are required for the design of such a fabrick, if it is set up as modestly as possible?

§35 For the academic instruction in technology to be completely comprehensible and of continuous use for those who are studying, 4) the instructor must necessarily demonstrate for every handcraft that he describes the primary and secondary materials, models of both useful and suggested tools and machines, and also, when possible, models of sample wares, and of various fabricks, both good and flawed ones, – if in the end a handcraft has been described in the lectures completely and precisely, only then does the instructor have to bring the auditors into the workshop itself, and see to it that they are shown all of the individual steps of work, as well as how these steps follow one another.

§36 Following this it is also 5) of great use in lectures on technology to include interesting reports on the invention, the general improvement, and the other fates of the arts: although including such reports, which do not actually belong to technology, in the instruction book itself would make it too long and expensive.

§37 Finally 6) it is the duty of the academic instructor to make both general technological writings as well as particular ones on individual parts and doctrines known to the one who is studying, so that he can subsequently use the best ones for the acquisition of diverse knowledge, and will not be misled by bad writings. So that no loss of time and mistakes should result from the dictation of book titles, both at the beginning of the general section, as well as before all individual parts and doctrines, the particular writings have to be listed completely, and only the more recent arrivals mentioned in addition.

§38 Finally, whoever who wants to gain a thorough knowledge of handcrafts must first, before proceeding to this study, learn the auxiliary sciences of technology, which the encyclopedia and methodology of the cameral sciences already made known to him right at the beginning, as he began his academic course of study. Then, however, he must choose that instructor who presents technology according to the determined rules, dutifully visit the lectures and workshops where possible, make drawings of the tools and machines which he becomes acquainted with, read the writings that are introduced to him by the instructor, and after making it through these academic years, take advantage of every opportunity to visit the workshops of the artisans.

Sixth Chapter: On the Use of Technology

§39 The use of technology reveals itself *firstly* with regard to the advancement of the general good, and *secondly*, with regard to the particular good of various kinds of citizens of a state.

§40 Justifying the general use of technology for the state does not depend on introducing evidence of the great influence of the flourishing of the handcrafts in the state for the general good, whose possibility and validity no one can doubt, but rather on showing to what degree technological study influences the thriving condition of the handcrafts, and therefore how such study advances the general wellbeing of all citizens.

§41 Experience teaches that chance certainly brought most inventions to light, but one should thank the technologist many times over that through the search for the rules and principles whereupon these inventions rest, which were mostly unknown to the inventor, and through the general publication of them, many were preserved from loss, and others were brought more quickly to completion and usefully expanded. Various important inventions for processing raw materials have their origin only in the industrious experiments and observations of those technologists equipped with the required knowledge of it, which would never occur to common handcraftsmen.

§42 The precise familiarity, not only with one handcraft, but with all of them, has often given rise to the opportunity of making workable suggestions for applying the methods, tools, and machines of one handcraft towards the improvement of another. Furthermore, one has technology to thank for the many transferals of advantageous establishments and kinds of methods of handcrafts from foreign lands to domestic ones. Even more so, since through all this technology promotes the flourishing of the handcrafts by furnishing, with thorough knowledge, those people whose duty it will one day be, as cameralists, to promote the flourishing of handcrafts in the state through appropriate laws, institutions, and individual ordinances.

§43 Good technological knowledge guarantees particular advantages, however, for the following individual kinds of citizens, such as 1) the husbandman (*Landwirth*), in part, inasmuch as he is with such complete knowledge in the position to pursue advantageous secondary trades, and in part because he can often give the acquired products, whose use he also knows, a more advantageous condition for use and can also in the meantime sell them more dearly, and in part, finally, in as much as in many cases he is capable of finding the right kind of buyers for his products, who can best use them, and hence pay the most for them.

§44 Then, 2) the *merchant*. If this individual, with regard to a ware with which he deals, knows exactly from which materials it is prepared, in what way, and according to which principles, he will thus be a better judge of its quality for bargaining. The signs of poor wares and their frequently hidden defects can certainly be evident to such an individual. He cannot be tricked by any falsifying or concealing of the defects. If he deals with raw materials, and if their use and the qualities that make them appropriate for processing are known to him, then here, too, he will better understand their purchase and will see afterwards, through good wares, the number of his customers increased. In many ways, the merchant equipped with thorough technological knowledge will also guide the hand of the fabricant so as best to support the turnover of goods.

§45 *Jurists* will be 3) in many cases in a better position to fulfill the duties of their offices, and will run into fewer unwelcome difficulties, if they are familiar with all the technical terms of handcraftsmen found in acts, trade documents, and other documents relative to handcrafts, and the designations for their works and tools. For the arbitration of disputes of closely related guilds, for the defense or contestation of their rights, they should not first learn with great difficulty in office what they already could learn scientifically, and more easily, pleasantly, and thoroughly.

§46 But finally, and above all, 4) the study of technology is indispensable for the future *cameralist*. It behooves him to promote the flourishing of the handcrafts, manufactures, and fabricks in the state, to contribute to the advantageous introduction of new ones that are still lacking, to judge whether the petitions for sovereign support for those fabricants who request it can proceed, and to guard against squandering public funds and the oppression of public servants. How can he dutifully accomplish all this without thorough knowledge of handcrafts? The determination of taxes, both for wages and for goods, also requires technological knowledge. Often, fabricks belong to the state itself, or they will be maintained under the ministration of one of the chambers of accounts, until an entrepreneur is found so that they do not completely close down, and leave so many people without food, and then they stand within the immediate oversight of the cameralists. Finally, there are found, under the authority of sovereign domain offices (*landesherrlichen Domainenämtern*), breweries, distilleries, tileworks, oil mills, etc., and, from these, writs should be executed and productivity increased, – how necessary to the camaralist is the knowledge of handcrafts, manufactures, and fabricks in all of these aforementioned cases, as well as others, not only what has been gained in the workshops, but also fundamental scientific knowledge!

Second Part: Literary History of Technology

§47 After the completed preparatory instruction in technology, its literary history now follows, which does not provide instruction in the invention of the arts and handcrafts, and their general improvements and developments, but rather shows, already in the *first* chapter, if, how, and where one first began to work on their theory, and how the science called technology came to be, and what its fate has been up to the present day, and then in the *second* chapter conveys knowledge of the general writings about this science.

First Chapter: History of Technology

§48 In former times one did not think it worth the effort, and perhaps not even proper, for scholars to describe handcrafts. Individual reports and remnants from antiquity teach us that the Greeks, Romans, and other nations practiced arts unknown to us, and achieved a high degree of perfection in many others which we also have. Meanwhile, we have in fact lost much, in that they did not leave behind any descriptions of their handcrafts for us.

§49 Already prior to the invention of the art of book printing there are manuscripts that contained descriptions of handcrafts and arts. Of these, various ones have been printed again which, however, are of no particular use. In the sixteenth century, various writers attempted descriptions of handworks and their secrets and published them under the title *Books of the Arts* (*Kunstbücher*). Many of those are only known to us by title. The ones passed down to us certainly contain several good contributions to the history of inventions, but for technology itself they are, as one can easily imagine, quite dispensable. More useful for us were the efforts by several engravers and painters in the seventeenth century, who drew the workshops, tools, and works of the handcraftsmen, and published them with descriptions.

§50 In the history of technology in the seventeenth century, Paul Jacob Marperger[9] was, to a certain degree, epoch-making for being the first, with great industry and precision, to describe individual handcrafts. He attracted the attention of both his contemporaries and followers, even those who were not actually handcraftsmen by trade, to what is pleasant and useful in this knowledge.

§51 After the first quarter of the eighteenth century, not only did more people in Germany begin to provide descriptions of handcrafts in dictionaries and other kinds of writings, but also – what was more important for technology – great scholars and entire learned societies chose the history of nature, physics, chemistry and mathematics, the supporting pillars of technology, as their favorite occupations.

§52 Since rulers finally also began to see how necessary and useful it is to promote the flourishing of handcrafts, manufactures and fabricks, as well as trades in general, and that for the attainment of this goal the members of the chamber of accounts must possess thorough knowledge of the handcrafts, thus individual instructors were appointed in these subjects, as well as the sciences of economics, civil regulation, and cameralism at universities, and in Halle first of all, which had great influence on the spread and perfection of technology.

§53 For most important contributions to technology, we have without dispute two excellent works of the French to thank. The *first* is the *Encyclopédie* described below. This dictionary, to be sure, does not just possess technological content, but is rather dedicated to objects of many other sciences. Since, however, the technological articles belong to those which are most thoroughly worked out, and since the descriptions are illustrated with several volumes of engravings, this encyclopedia, whose first author was Diderot, and on which

9 Paul Jacob Marperger (1656–1730) was a German writer known for his works on cameralism and economic matters.

twenty of the most famous men in France – a certain Montesquieu, Voltaire, d'Alembert, du Morveau, etc. – have worked, thusly belongs to the most important technological writings.

§54 Almost even more honor is afforded to the French by the detailed descriptions of handcrafts and arts that the members of the Parisian *Academy of Sciences* have in part worked out on their own, and in part have had famous men develop under their supervision, which bear the general title, *Descriptions des arts et métiers* (Descriptions of the arts and trades). Although the academy, already founded at Colbert's[10] initiative in 1666, had already drafted the plan for such a generally useful work in the first years after its inception, hindrances of various kinds did not allow for the beginning to be made with the imprint of the descriptions before the year 1761, and since this time they have been continued with remarkable eagerness. Every handcraft comprises one volume in large folio with an individual title, and the descriptions are accompanied by the most excellent engravings, which illustrate not only the set-up of the workshops, tools, and machines, but also even techniques (*Handgriffe*) and crafts (*Kunstgriffe*). An index of all volumes published up through 1775 is found in Beckmann's library, volume 7, page 43.

§55 In Germany in the year 1762, an initial translation of this excellent, but expensive work, was begun, which bore the general title, *Schauplatz der Künste und Handwerke* (Theater of the arts and handcrafts). Von Justus[11] took care of the translation of the first four volumes, and afterwards (Daniel) Gottfried Schreber[12] continued through the thirteenth volume. With additional notations, the German edition has the advantage of the original manuscript, but it also falls slightly behind the French original in the clarity and elegance of the copied engravings. A table of all translated handworks and arts can be found in Beckmann's *Library*, volume 9, page 504, and it is to be desired that this translation could be continued in similar quality.[13]

§56 From the general bibliography, which is to be conveyed in the following paragraph, and from the particular writings leading through the entire system of technology, will be clear how very exquisitely the Germans in recent times have made use of them in their work on this science, and the classical

10 Jean-Baptiste Colbert (1719–1783), French statesman.

11 Johann Heinrich Gottlob von Justi (1717–1771), German political economist.

12 Daniel Gottfried Schreber (1708–1777), German political economist.

13 Johann Beckmann, *Physikalisch-ökonomische Bibliothek* (Göttingen, 1770–1804). The ninth volume was published in 1778.

technological writers deserve to be named here, primarily: Zink,[14] von Justus,[15] Schreber, von Pfeiffer,[16] Halle,[17] Bergius,[18] Sprengel,[19] Beckmann, Jacobsson, Jung,[20] and many others.

§57 At universities, the theory of handcrafts was initially only presented in a section of the cameral system. Already some ten years ago, however, Privy Councilor and Professor Johann Beckmann in Göttingen earned the honor of first presenting particular lectures on this science, to which he gave the well-chosen name technology, in accordance with a textbook written for this purpose, and his excellent example has now been followed by many instructors at universities. Technological instruction has been introduced to great advantage in a few schools, and for this to become a more general practice is to be desired, particularly in such schools which are attended by future handcraftsmen.

§58 Primarily, however, the theory of handcrafts has gained much in perspective and thoroughness in recent times, so that one has begun to think through the theory of nature, chemistry, and mathematics in more particular application to handcrafts. Primarily, one has for this the efforts of Hoffmann,[21] Weigel,[22] Wiegleb,[23] Achard,[24] Erxleben,[25] Cartheuser,[26] Weber,[27] Model,[28] Crell,[29] Suckow,[30] Karsten,[31] Gmelin,[32] and many others to thank.

14 Georg Heinrich Zincke (1692–1769), German jurist and cameralist.
15 Johann Heinrich Gottlob von Justi (1717–1771), German political economist.
16 Johann Friedrich von Pfeiffer (1718–1787), German cameralist.
17 Johann Samuel Halle (1727–1810), Prussian historian.
18 Johann Heinrich Ludwig Bergius (1718–1781), German economist.
19 Peter Nathanael Sprengel (1737–1814), German historian.
20 Johann Heinrich Jung (1740–1817), German doctor and economist.
21 Carl August Hoffmann (1756–1833), German apothecary.
22 Christian Ehrenfried Weigel (1747–1831), German professor of chemistry and other subjects at Greifswald.
23 Johann Christian Wiegleb (1732–1800), German apothecary and chemist.
24 Franz Carl Achard (1753–1821), German chemist and geoscientist.
25 Johann Christian Polycarp Erxleben (1744–1777), German naturalist.
26 Friedrich A. Cartheuser (1734–1796), German geologist and minerologist.
27 Friedrich Benedict Weber (1774–1848), German cameralist.
28 Johann Georg Model (1711–1775), German chemist.
29 Lorenz Florenz Friedrich von Crell (1744–1816), German chemist.
30 Georg Adolf Suckow (1751–1813), German professor of chemistry, physics, mining, and minerology.
31 Franz Christian Lorenz Karsten (1751–1829), German economist.
32 Johann Friedrich Gmelin (1748–1804), German chemist.

Second Chapter: The Bibliography

§59 The bibliography, as the second section of the literary history, imparts a knowledge of technological writings according to certain classes. Its main division is into the general and the specific, based on whether its contents either are *not* directed toward an individual part or individual material of technology, or if simply an entire part or individual material of technology is treated therein. Only the index to the former belongs here, whereas the particular technological writings are themselves listed within the individual parts of technology.

§60 The general technological writings, however, are to be listed according to the following classes:

1) The *libraries*, which give notice of all the writings wherein handworks are treated, and which a) either merely make known the book titles, or also simultaneously include judgements on the quality and usefulness of the books, b) are either periodical works, which are published in part, gradually, by the publisher, or not. Of these such works which touch lightly upon technology are not included. Here, one can only mention those in which, beyond the writings of other sciences, the technological writings are demonstrated as well.

§61

2) *Dictionaries*, wherein the parts of technology and its technical terms are explained in alphabetical order, and which again are either exclusively of technological content or have also spread into other sciences.

§62

3) *Systems* and *instruction books* wherein at least most, if not all, handcrafts are described, each in connection to the other, with or without order, extensively or in short sentences. They are a) either such works, in which technology is presented in connection with economy and the sciences of governance and finance, or such, wherein technology, separated from those sciences closely related to it, is treated b) according to the intention of the author either as a guide for oral instruction, or not.

§63

4) Various writings, collections, magazines, in which individual parts or doctrines of technology are either presented alone, or – which is commonly the case – mixed with doctrines of other sciences which, firstly, are either published by learned societies, or not and, secondly, are either periodical works, or not.

§64

5) Finally, the *historical-statistical works* complete the list, wherein reports are shared on the invention, the general improvements, and the current state and condition of handcrafts, manufactures, and fabricks, which are currently found in various countries.

I Libraries[33]

§65

1) Zincke, Georg Heinrich. *Cameralisten-Bibliothek* (Library for cameralists). Parts 1–4. Leipzig 1751–2. 8.
2) Moser, Johann Jacob. *Bibliothek von Oeconomischen-Cameral-Policey-Handlungs-Manufactur-Mechanischen und Bergwerks Gesetzten Schriften und kleinen Abhandlungen* [Library of legal writings and small treatises pertaining to the economical, cameralist, governance, trade, manufacture, the mechanical, and mining]. Ulm 1758. 8.
3) Neuenhahn, Carl Ludewig. *Vermischte Bibliothek, oder Auszüge aus verschiedenen zur Arzneigelahrtheit, Chymie, Naturkunde, Oekonomie, zu Manufacturen und Künsten gehörenden akademischen Streitschriften und Abhandlungen* [Mixed library, or excerpts from various academic polemics and treatises relating to medicine, chemistry, the study of nature, economics, manufactures and the arts]. Braunschweig 1758.8.
4) Bergius, Johann Heinrich Ludwig. *Cameralisten-Bibliothek, oder vollständiges Verzeichnis derjenigen Bücher, Schriften und Abhandlungen, welche von dem Oeconomie- Policey-Finanz- und Cameralwesen und verschiedenen andern damit verbundenen Wissenschaften, auch von der dahin einschlagenden Rechtsgelehrsamkeit handeln* [Library for cameralists, or complete index of those books, writings and treatises which concern matters of economics, governance, finance, and cameralism and various other related sciences, and also pertaining juridical matters.] Preface, Christian Ernst von Windheims. Nuremberg 1762. 8.

---

33 The following method has been used for providing bibliographic information: the entries correspond to the editions listed in Lamprecht's text. Additional bibliographic details have been provided to make entries more complete. This includes the full names of individual authors and replacing abbreviations with the corresponding words. English translations of the titles are provided following the foreign language original. The final number in each of Lamprecht's entries refers to the size of the original publication (8 = octavo, 4 = quarto, etc.).

5) *Allgemeine deutsche Bibliothek.* [General German library]. Berlin und Stettin. Since 1766. Published by Friedrich Nicolai, 8.
6) Johann Beckmann, *Physicalisch-ökonomische Bibliothek, worinn von den neuesten Büchern, welche die Naturgeschichte, Naturlehre und die Land- und Stadtwirtschaft betreffen, zuverläßige und vollständige Nachrichten ertheilet werden* [Physical-economical library, wherein information is provided concerning the latest books having to do with the study of nature, agriculture, and town economy]. Göttingen, since 1770. 8.
7) Johann Traugott Müller, *Einleitung in die ökonomische und physikalische Bücherkunde und die damit verbundenen Wissenschaften bis auf die neuesten Zeiten* [Introduction into the economical and physical bibliography and related sciences, up to the most recent times]. Leipzig 1780, 1782. 2 vols. 8.
8) Georg Rudolph Böhmer, *Bibliotheca scriptorum historiae naturalis oeconomiae aliarumque artium ac scientiarum* [Library of writings in natural history, economics, and other arts and sciences]. Leipzig 1785, 1786. 2 vols. 8.
9) Johann Hermann Pfingsten, *Bibliothek ausländischer Chemisten, Mineralogen und mit Mineralien beschäftigter Fabrikanten nebst derley biographischen Nachrichten* [Library of foreign chemists, minerologists, and fabricants working with the minerals, along with biographical reports about them]. Nuremberg 1781, 1782. 2 vols. 8.
10) *Allgemeine Litteratur-Zeitung* [General literature periodical], since 1785.

II Dictionaries

a) *German*

§66

1) Adrian Beier, *Allgemeines Handlungs- Kunst- Berg- und Handwercks-Lexicon* [General lexicon of trade, arts, mining, and handcraft] Jena 1722. 4.
2) Johann Hübner, *Kurieuses und reales Natur- Kunst- Berg- Gewerck- und Handlungslexikon* [Strange and real lexicon of nature, the arts, mining, industry, and trade]. Leipzig 1762. 4.
3) Johann Heinrich Zedler, *Großes und vollständiges Universallexikon aller Künste und Wissenschaften.* [Large and complete universal lexicon of all arts and sciences]. Leipzig 1731f. Fol.
4) George Heinrich Zinke, *Teutsches Real- Manufaktur- und Handlungs-Lexicon* [German royal manufacture and trade lexicon]. Leipzig 1745. Vol. 1.
5) Johann Samuel Heinsius. *Allgemeine Schatz-Kammer der Kauffmannschafft, oder vollständiges Lexicon Aller Handlung und Gewerbe* [General

treasure-room for merchants, or complete lexicon of all trade and industry]. Leipzig 1741–1743. 5 vols. Fol.

6) Johann Theodor Jablonsky, *Allgemeines Lexikon der Künste und Wissenschaften* [General lexicon of the arts and sciences]. Leipzig 1721. 4.

7) Carl Günter Ludovici. *Eröffnete Akademie der Kaufleute oder vollständiges Kaufmanns-Lexicon.* [Open academy for merchants or complete merchants' lexicon]. Leipzig 1752–1756. 5 vols. 8.

8) Johann Heinrich Ludwig Bergius. *Polizey- und Cameral-Magazin* [Magazine for governance and cameralism] Frankfurt am Mayn since 1767–1774. 9 vols. 4. And *Neues Polizey- und Cameral-Magazin* [New magazine for governance and cameralism] Leipzig 1775–1780. 6 vols. 4.

9) Johann Georg Krünitz. *Ökonomisch-technologische Encyklopädie, oder allgemeines System der Staats- Stadt- Haus- und Landwirthschaft, in alphabetischer Ordnung* [Economic-technological encyclopedia, or general system of state, civic, household and agricultural economy, in alphabetical order]. Berlin since 1773. 37 vols. 8.

10) *Deutsche Encyclopädie, oder allgemeines Real-Wörterbuch aller Künste und Wissenschaften* [German encyclopedia, or general subject dictionary of all arts and sciences]. By a learned society. Frankfurt am Mayn since 1778. 8 vols. 4.

11) Johann Georg Krünitz et al. *Auszug aus Krünitz ökonomisch-technologischer Encyklopädie* [Excerpt from Krünitz's *economical-technological encyclopedia*]. Berlin since 1786. 8.

12) Johan Karl Gottfried Jacobsson, *Technologisches Wörterbuch* [Technological dictionary] Berlin 1781, 1783. 4 vols. 4.

13) Johann Georg Meusel, *Teutsches Künstlerlexikon oder Verzeichnis der jetzt lebenden deutschen Künstler.* [German lexicon of artisans or index of German artisans alive today]. Lemgo 1778. 8.

14) Johann Heinrich Fueßli, *Allgemeines Künstlerlexikon* [General lexicon of artisans]. Zürich, most recent edition, 1779. Fol.

15) Georg Gottfried Strelin, *Realwörterbuch für Kameralisten und Oekonomen* [Subject dictionary for cameralists and economists]. Since 1783. 3 vols. 8.

b) *Foreign Languages*

§67

1) *Dictionnaire universel de la langue francoise (sic) des arts et des sciences* [Universal dictionary of the French language of the arts and sciences]. By members of the French Academy. Paris 1694–1740. fol.

2) Ephraim Chambers. *Cyclopaedia, or an universal Dictionary of Arts and Sciences*. London 1742. fol. Vol. I and II, and *Supplement to Mr. Chambers Cyclopaedia*. London 1753. Vols. I. and II.

3) Thomas Dyche (trans.). *Nouveau Dictionnaire universel des arts et des sciences* [New universal dictionary of the arts and the sciences]. Translated from the English. Avignon 1753. 4.

4) Gianfrancesco Pivati. *Nuovo dizionario scientifico e curioso, sacro-profano* [New scientific, curious, sacred, and profane dictionary]. Venice 1746–1757. 10 vols. Folio.

5) Dénis Diderot and Jean d'Alembert. *Encyclopaedia* (sic) *ou Dictionnaire raisonné des Sciences des Arts et Metiers etc.* [Encyclopedia or reasonable dictionary of the sciences, arts, and trades, etc.] Paris 1751–1765. Vols. I–XVII folio as well as 12 volumes of engravings and four supplemental volumes. Amsterdam 1776–1777. Among several new editions of this dictionary, the most notable is the one, which appeared under the title, *Encyclopedie ou Dictionnaire universel raisonné des connoissances humaines, mis en ordre par de Felice* [Encyclopedia or reasonable universal dictionary of human knowledge, put in order by Fortunato Bartolomeo de Felice], Yverdon 1770–1775. 42 vols., 10 vols. of engravings. 4.

6) *Table analytique des matieres continues dans les 23 Vol. in fol. du Dictionnaire – et dans son Supplement* [Analytical table of materials contained in the 23 folio volumes of the dictionary and its supplement] Paris 1780. Vol. II.

7) *Table analytique et raisonnée des materieres continues dans les 29. Vol. in 4. du Dictionnaire des sciences, des arts* [Analytical and reasonable table of materials contained in the 29 quarto volumes of the Dictionary of the Sciences and the Arts]. Geneva 1780–1783. Vols. I–IV. 4.

8) Thomas Dyche. *A new and complete Dictionary of arts and sciences.* London 1754. Vols. I and II. 8.

9) Temple Henry Croker. *The complete Dictionary of Arts and Sciences.* London 1768, 1769. Vols. I–III. Folio.

10) Philippe Macquer. *Dictionnaire portatif des arts et metiers* [Portable dictionary of the arts and sciences]. Paris 1766. Vols. I and II. 12.

11) Louis-Abel de Bonafous de Fontenay (Abbé). *Dictionnaire des Artistes.* Paris 1777. Vols. I and II.

12) Jean-Marie Roland de la Platière. *Encyclopedie* [sic] *methodique des manufactures, arts et metiers* [Methodical encyclopedia of manufactures, arts, and trades]. Paris 1784–1785. Vols. I and II.

13) Macquer, *Chemisches Wörterbuch*, translated by Leonhardi. Leipzig 1781, 1782. 2 parts. 8.

III Systems and Instruction Books (*Lehrbücher*)

a) *In Connection with Economy, and the Sciences of Governance and Finance*

§68

1) George Heinrich Zinke, *Grundriß einer Einleitung zu den Kameralwissenschaften* [Outline of an introduction to the cameral sciences]. Leipzig 1742. 8.
2) George Heinrich Zinke, *Anfangsgründe der Kameralwissenschaften* [Beginning principles of the cameral-sciences]. Leipzig 1755. 2 vols. 8.
3) Lorenz Johann Daniel Suckow, *Die Kameralwissenschaften nach dem Grund-Risse des Herrn Geheimen-Rath Darjes*. Jena 1768. 8.
4) Johann Christoph Adelung. *Kurzer Begriff menschlicher Fertigkeiten und Kenntnisse, in so fern sie auf Erwerb des Unterhalts, auf Vergnügen, auf Wissenschaft, und auf Regierung der Gesellschaft abzielen* [Short outline of human talents and knowledge, as related to earning a wage, for pleasure, for science, and for governing society]. Leipzig 1778–1781. 4 vols. 8.
5) Johann Friedrich von Pfeiffer, *Lehrbegriff sämtlicher ökonomischen und Kameralwissenschaften* [Outline of collected economical and cameral sciences]. Mannheim 1764–1772. 4 vols. along with a supplement in 4.
6) Johann Christian Förster, *Einleitung in die Kameral- Polizey- und Finanzwissenschaften* [Introduction to the sciences of cameralism, governance, and finance]. Halle 1771. 8.
7) Johann Christian Förster, *Entwurf der Land- Stadt- und Staatswirtschaft* [Plan of the agrarian, civic, state economy]. Berlin 1782. 8.
8) Johann Friedrich von Pfeifer, *Grundsätze der Universalkameralwissenschaften* [Principles of the universal cameral sciences]. Frankfurt am Main 1783. 2 vols. 8.
9) Georg Friedrich von Lamprecht, *Encyklopädie und Methodologie der ökonomischen, politischen und Kameralwissenschaften* [Encyclopedia and methodology of the economic, political, and cameral sciences]. Halle 1785. 8.

b) *Individual Technological Systems and Instruction Books*

§69

1) *Eigentliche Beschreibung aller Stände auff Erden, allen Ständen zu Ehren* [Proper description of all worldly estates, in honor of all estates]. Siegmund Feyerabend. Frankfurth am Mayn 1568.
2) Tommaso Garzoni, *Piazza universale, das ist: allgemeiner Schauplatz, oder Markt und Zusammenkunft aller Professionen, Künsten, Geschäfften, Händlen* [sic] *und Handwercken so in der ganzen Welt geübt werden* [Universal plaza, that is: arena or market and meeting of all professions, arts,

businesses, trades and handcrafts practiced in the entire world]. Frankfurt am Mayn 1619.

3) Christoph Weigel, *Abbildung und Beschreibung der Gemein Nützlichen Hauptstände von denen Regenten und ihren so in Friedens- als Kriegs-Zeiten zugeordneten Bedienten an, biß auf alle Künstler und Handwercker.* [Illustration and description of the primary commonly useful occupations, from the regents and their servants in times of peace and war, down to all artisans and handcraftsmen]. Regensburg 1698. 8.

4) *Spectaculum Naturae et Artium, in vier Sprachen, Deutsch, Lateinisch, Französisch und Italiänisch* [Spectacle of nature and art in four languages, German, Latin, French, and Italian]. Berlin 1761–1765. 4.

5) Members of the royal academy of sciences. *Descriptions des arts et metiers* [Descriptions of arts and trades], Paris 1761f. 47 vols. Folio.
Of this, a new edition under the title:

6) *Descriptions des arts et metiers* [Descriptions of arts and trades]. Neufchatel 1771–1782. 18 vols.
An attempted (incomplete) German translation bears the title:

7) *Schauplatz der Künste und Handwerker, oder vollständige Beschreibung derselben, verfertiget oder gebilliget von denen Herrn der Akademie der Wissenschaften zu Paris* [Arena of the arts and handcrafts, or complete description of them, composed or approved by the gentlemen of the academy of sciences in Paris]. Translated by Johann Heinrich Gottlob von Justi. Vols. 1–4. Berlin and Leipzig 1762–1765, and finally by Johann Christian Harrnpeter, vol. 15. Nuremberg and Leipzig 1782.

8) Johann Heinrich Gottlob von Justi, *Abhandlung von Manufakturen und Fabriken* [Treatise on manufactures and fabricks]. 2nd edition by Johann Beckmann. Berlin 1780. 2 vols. 8.

9) Johann Samuel Halle, *Werkstätte der heutigen Künste, oder die neue Kunsthistorie* [Workplace of contemporary arts, or the new history of the arts]. Brandenburg and Leipzig 1761–1779. 6 vols.

10) Peter Nathanael Sprengel, *Handwerke in Tabellen, fortgesetzt von Hartwig* [Handcrafts in tables, continued by Otto Ludwig Hartwig]. Berlin 1767–1777. 15 editions. 8. New imprint of the first edition. Berlin 1778–1782. 8.

11) Johann Beckmann, *Anleitung zur Technologie oder zur Kenntniß der Handwerke, Fabriken und Manufakturen* [Guide to technology or to knowledge of the handcrafts, fabricks, and manufactures]. Göttingen 1780. 8.

12) Johann Samuel Halle *Technologie oder die mechanischen Künste, als ein vermehrter Auszug aus der neuen Kunsthistorie* [Technology or the mechanical arts, as an augmented excerpt from the recent history of the arts]. Brandenburg 1782. 8.

13) Johann Friedrich von Pfeifer, *Die Manufakturen und Fabriken Deutschlands nach ihrer heutigen Lage betrachtet und mit Vorschlägen zu ihrer Verbesserung begleitet* [The manufactures and fabricks of Germany observed according to their current situation and accompanied by suggestions for their improvement]. Frankfurt am Main 1780–1. 2 vols.
14) Johann Heinrich Jung-Stilling, *Versuch eines Lehrbuchs der Fabrikwissenschaft zum Gebrauch akademischer Vorlesungen* [Attempt at a textbook for the science of fabricks for the use of academic lectures]. Nuremberg 1785.
15) Cunradi, *Anleitung zum Studium der Technologie, oder kurze und faßliche Beschreibung verschiedener Künste und Handwerke* [Guide for the study of technology, or short and comprehensive description of various arts and handcrafts]. Leipzig 1786.

IV Miscellaneous/Various Writings, Collections, Magazines

a) *Works of Learned Societies*

§70

1) *Philosophical Transactions: Giving Some Account of the Present Undertakings, Studies and Labours of the Ingeniuous in Many Parts of the World.* London 1685f.
2) *Histoire de l'academie royale des sciences depuis l'année 1699 jusqu'en 1759, avec les memoires de mathematique et de physique pour les mêmes années* [History of the royal academy of sciences from the year 1699 through 1759, with the memoirs of mathematics and physics for the same years]. Paris 1699–1780. 4.
3) *Histoire de l'academie royale des sciences a Berlin – avec les memoires* [History of the royal academy of sciences in Berlin, with memoirs]. Berlin 1746–1770. Vol. 1–25. 4; *nouveaux memoires* (new memoirs) 1770–1786. Vols. 1–15. 4.
4) *scientiarum et artium instituto atque Academia Commentarii* [Institute of the sciences and the arts and academic commentaries] Bologna 1731–1757. Vols. I–VIII. Translated by Leske. Vols. 1 and 2. Brandenburg.
5) *Commentarii academiae scientiarum imperialis Petropolitanae*. [Commentaries of the Petropolis imperial academy of sciences] 1726–1752. Vols. I–XIV. 4. new commentaries. Petropolis. 1750–1774. Vols. 1–18. and acts of the imperial academy of sciences for the years 1777–1786.
6) Könglig Svenska Wetenskaps Akademiens Handlingar [Procedings of the royal Swedish academy of science], Stockholm 1740–1779. Vols. I–XL. 8. Könglig Wetenskaps Akademiens nya Handlingar (New procedings of the royal academy of science). Stockholm 1780. ––. Der königl. schwedischen Akademie der Wissenscahften Abhandlungen aus der Naturlehre,

Haushaltungskunst [The royal Swedish academy of sciences procedings from the study of nature, economics]. Translated by Kästner. Hamburg und Leipzig, since 1749–1783. Vols. I–XL and *Neue Abhandlungen* [New treatises] 1784. Vol. I.

7) *Versuche und Abhandlungen der Naturforschenden Gesellschaft in Danzig* [Experiments and treatises of the society for natural research in Danzig] 1747–1758. Vols. 1–3. And new collection of experiments, Danzig 1778. Vol. 1.

8) *Commentarii societ. reg. scient. Götting.* [Commentaries of the royal scientific academy of Göttingen] 1751. 1752. Tom. I–IV. New commentaries of the royal scientific academy of Göttingen. 1769–1777. Vols. I–VIII. *Commentationes soc. reg. Sc. Göttingen* [Commentaries of the royal scientific academy of Göttingen] 1779. Vols. I–VI. 4.

9) *Acta academiae electoralis Moguntinae scientiarum utilium quae Erfordiae est.* [Academic acts of the electorate of Main which is useful for sciences in Erfort]. Erfurt and Göttingen 1757–1776. Vols. I–III.

10) *Vermischte Abhandlungen der physisch-chemischen Warschauer Gesellschaft zur Befördering der Naturkunde, Oekonomie, Manufakturen u.s.w.* [Various treatises of the physical-chemical society of Warsaw for the advancement of the study of nature, economics, manufacture] Warschau und Dresden, 1768. Vol. 1.

11) *Bemerkungen der churpfälzischen physikalisch ökonomischen Gesellschaft* [Remarks of the Electoral Palatine physical economical society]. Mannheim since 1770.

12) *Beschäftigungen der Berliner Gesellschaft Naturforschender Freunde* [Preoccupations of the Berlin society of nature-researching friends]. Berlin 1775–1779. Vols. 1 to 4. und *Schriften der Berlinischen Gesellschaft Naturforschender Freunde* [Writings of the Berlin society for nature-researching friends]. Berlin since 1780.

13) *Sammlungen nützlicher Unterrichte, herausgegeben von der Kaiserl. Königl. Gesellschaft des Ackerbaues und nützlicher Künste in Crain* [Collections of useful instruction, published by the Imperial Royal Society of agricluture and useful arts in Crain]. Laybach 1773–1779. Collection 1 to 4.

14) *Oekonomische Nachrichten der patriotischen Gesellschaft in Schlesien* [Economic reports of the patriotic society in Schlesia]. Breslau 1773–1779. Vols. 1–7. *Neue ökonom. Nachrichten aufs Jahr 1780* [New economic reports for the year 1780]. Vol. 1.

15) *Verhandelingen van het Bataviasch Gewootschaf der Konsten en Wetenschappen* [Treatises of the Batavian society of arts and sciences]. Part one. Batavia 1779. 8.

16) *Memoires de la societé établie à Geneve pour l'encouragement des arts et de l'agriculture* [Memoirs of the society established in Geneva for the promotion of the arts and of agriculture]. Genève 1780. Vol. 1, parts 1 and 2.

17) *Kleine Schriften der Leipziger ökonomischen Societät* [Short writings of the Leipzig Economical Society]. Friedrichstadt, 1778. And *Schriften der Leipziger Societät* [Writings of the Leipzig Society] 1771–1781. Vols. 1–4. 8.

b) *Works of Individual Scholars, Namely*

a) Such, in Which Technological and Economical Treatises on Trade, Governance, and Cameralism Are Found Together

§71

1) Christian Wolff, *Allerhand nützliche Versuche, dadurch zu genauer Erkäntniß der Natur und Kunst der Weg gebähnet wird denen Liebhabern der Wahrheit* [Assorted useful experiments though which the way to precise knowledge of nature and art will is paved for the lovers of truth]. Halle 1721–1723. Vols. 1–3.

2) George Heinrich Zinke, *Leipziger Sammlungen von wirthschaftlichen Polizey-Kammer- und Finanzsachen* [Leipzig collections of economic matters of governance, chamber, and finance]. Leipzig 1746–1767. Vols. 1–16. 8.

3) *Hannöverische Sammlungen kleiner Ausführungen aus verschiedenen Wissenschaften* [Hannoverian collections of short accomplishments from various sciences]. Hannover 1750–1754. Vols. 1–4. 4.

4) *Hannöversche nützliche Sammlungen* [Hannoverian useful collections]. Hannover 1755–1758. 4 vols. 4.

5) *Hannöverische Beyträge zum Nutzen und Vergnügen* [Hannoverian contributions for use and pleasure]. Hannover 1759–1762. 4 vols. 4.

6) *Hannöverisches Magazin* [Hannoverian magazine], one volume annually since 1763. 4.

7) *Oekonomische Nachrichten* [Economic reports]. Leipzig 1750–1763. 15 vols. 8.

8) *Neue ökonomische Nachrichten* [New economic reports]. Leipzig 1763–1773. 5 vols. 8.

9) *Danziger Erfahrungen zur Beförderung der Einsicht in die Natur und Kunst von Hanow* [Danzig experiences for the promotion of insight into the nature and art of Hanow]. 1750–1758. 4.

10) Daniel Gottfried Schreber, *Sammlungen verschiedener Schriften* [Collections of various writings]. Halle 1755–1765. 16 parts. 8.

11) Daniel Gottfried Schreber, *neue Sammlungen* [New collections]. Bützow und Wismar. 1762–1765. 8 parts. 8.

12) Daniel Gottfried Schreber, *neue Kameralschriften* [New writings on cameralism]. Halle und Leipzig. 1765–1769. 8 parts. 8.
13) *Hamburgisches Magazin*. Hamburg 1747–1763. 8.
14) *Neues hamburgisches Magazin*, since 1767. Vols. 1–19. 8.
15) Johann Heinrich Gottlob von Justi, *Göttingische Polizeyamtsnachrichten* [Göttingen reports from the civil governance office]. Göttingen 1755–1757. Vols. 1–3. 4.
16) Johann Heinrich Gottlob von Justi *Ökonomische Schriften über die wichtigsten Gegenstände der Stadt- und Landwirthschaft* [Economic writings on the most important matters of state and agrarian economy]. Berlin und Leipzig 1760. Vols. 1 and 2.
17) *Kopenhagner Magazin, von ökonomischen, Kameral-Manufaktur, Gesetzen, Schriften und kleinen Abhandlungen* [Kopenhagen magazine on economic- and cameral-manufacture-related laws, writings, and small treatises]. 1759–1764. Vols. 1–3.
18) *Journal économique, ou memoires sur les arts et l'agriculture* [Economic journal, or memoires on the arts and agriculture]. Paris 1751–1769.
19) *Allgemeines Magazin der Natur, Künste und Wissenschaften* [General magazine on nature, arts, and sciences]. Leipzig 1753–1767. Vols. 1–12. 8.
20) *Natur- und Kunstkabinet, oder Sammlung nützlicher Nachrichten zur Beförderung der Naturkunde und Manufakturen* [Cabinet for nature and art, or collection of useful reports for the advancement of the study of nature and manufactures]. Jena 1755. Vol. 1. 8.
21) *Bremisches Magazin zur Ausbreitung der Wissenschaften, Künste und Tugend* [Bremen magazine for the expansion of the sciences, arts, and virtue]. Hannover 1756–1765. Vols. 1–7. 8.
22) *Neues Bremisches Magazin*. Bremen 1766–1769. Vols. 1 and 2.
23) Johann Jakob Moser, *Schwäbische Nachrichten von Oekonomie-Manufaktur- mechanischen und Bergwerkssachen* [Schwabian reports on matters of economy, manufacture, mechanics, and mining]. Stuttgart 1757. 8.
24) *Oekonomische Schriften über die nöthigsten Gegenstände der Stadt- und Landwirthschaft* [Economic writings on the most necessary matters of state and agrarian economy]. Berlin 1760. Vols. 1 and 2. 8.
25) *Museum rusticum et commerciale* [Rustic and commercial museum], *or select papers on agriculture, commerce, arts and manufactures*. London 1763–1769. 8. *Auserlesene Schriften, den Ackerbau, Handlung, Künste und Manufakturen betreffend, aus dem Englischen von J. J. Schwab* [Selected writings relating to agriculture, trade, the arts and manufactures, from the English by J. J. Schwab]. Leipzig 1764. 8.

26) *Giornale d'Italia spettante alle scienza naturale et principalmente all'agricoltura, all'arte* [Journal of Italy concerning all natural sciences and principally agriculture and arts]. Venice 1764–1776. Vols. I–XII.

27) *Gemeinnütziger Vorrath auserlesener Aufsätze zur Beförderung der Haushaltungswissenschaft, Künste, Manufakturen, Fabriken* [Public supply of chosen essays from the advancement of household science, the arts, manufactures, fabricks]. Leipzig 1767–1768. Parts 1–3. 8.

28) *Berlinische Sammlungen zur Beförderung der Arzneywissenschaft, Naturgeschichte, Haushaltungskunst, Kameralwissenschaft und dahin einschlagenden Litteratur* [Berlin collections for the advancement of medicine, natural history, household management, cameralism, and related literature]. Berlin 1768–1777. Vols. 1–9.

29) Jean Étienne Guettard. *Memoires sur differentes parties des sciences et des arts* [Memoirs on differents parts of the sciences and the arts]. Paris. 1768–1770. Vols. 1–3.

30) *Gemeinnützige Nachrichten aus dem Reiche der Wissenschaften und Künste* [Public reports from the realm of the sciences and the arts]. Hamburg 1768–1769. 4.

31) *Mannigfaltigkeiten, eine gemeinnützige Wochenschrift* [Miscellanea, a public weekly paper]. Berlin 1770–1773. Vols. 1–4. *Neue Mannigfaltigkeiten* [New miscellanea]. Berlin 1774–1777. Vols. 1–4. *Neueste Mannigfaltigkeiten* [The most recent miscellanea]. Berlin 1778–1780. Vols. 1–3. and *Allerneueste Mannigfaltigkeiten* (The most up-to-date miscellanea) since 1781.

32) François Rozier. *Observations sur la physique, l'histoire naturelle et sur les arts, contenant l'abregé des mem. des academ. de Berlin, Londres, Bologne etc. depuis l'ann. 1770* [Observations on physics, natural history and the arts, containing the summary of memoirs from the academies of Berlin, London, Balogne, and others from the year 1770 onward]. Paris 1773–1786. Vols. 1–20.

33) *Sammlung brauchbarer Abhandlungen aus Rozier Betrachtungen über die Natur und Kunst* [Collection of useful treatises from Rozier's observations on nature and art]. Leipzig 1775–1776. Vols. 1 and 2.

34) *Bernisches Magazin der Natur, Künste und Wissenschaften* [Bernian magazine for nature, the arts, and sciences]. Bern 1778. Vols. 1 and 2.

35) *Gazette d'Agricultur, commerce, arts et finances, pour l'année 1775* [Gazette for agriculture, commerce, arts and finances for the year 1775]. Paris 1775. 4.

36) *Gothaisches Magazin der Künste und Wissenschaften.* [Gotha magazine for the arts and sciences]. Gotha 1776. Vol. 1.

37) *Diario economico d'agricultura, manufatture e commercio* [Economic journal for agriculture, manufacture, and commerce]. Rome 1776.

38) *Italiänische Bibliothek, oder Sammlungen der merkwürdigsten kleinen Abhandlungen zur Naturgeschichte, Oekonomie, und Fabrikwesen, aus den neuesten italiänischen Monatsschriften.* [Italian library, or collections of the most noteworthy small treatises for the history of nature, economics, and fabrick sciences, from the most recent Italian monthly periodicals]. Leipzig 1778–1779. Vols. 1 and 2.

39) Johann Beckmann, *Beckmanns Beyträge zur Oekonomie, Technologie, Polizey- und Kameralwissenschaft* [Beckmann's contributions to economy, technology, and the sciences of governance and cameralism], since 1779. Vols. 1–6.

40) *Frankfurther Beyträge zur Ausbreitung nützlichen Künste und Wissenschaften* [Frankfurt contributions to the expansion of useful arts and sciences], Frankfurt 1780.

41) Christian Joseph Jagemann, *Magazin der italiänischen Litteratur und Künste* [Magazine of Italian literature and the arts]. Weimar 1780. 8.

42) *Physikalische Brieftasche, oder Nachrichten aus dem Gebiet der Landwirthschaft, Naturgeschichte, Technologie, Physik und Chemie* [Physical attaché, or reports from the area of agriculture, natural history, technology, physics, and chemistry]. Vienna, beginning July 1, 1787, one edition weekly. 8.

43) *Berlinisches Magazin der Wissenschaften und Künste* [Berlin magazine of the sciences and the arts]. Berlin 1782. Vol. 1.

44) Reinhold Forster, *Sammlung von Abhandlungen ökonomischen und technologischen Inhalts* [Collection of treatises of economic and technological content]. Halle 1784. 8.

b) Such Works, in Which Are to Be Found Only Technological Doctrines or Chemical and Physical Theories Applied to Handcrafts and Arts

§72

1) *Natuur en Kunstkabinet, of Kabinet der natuuriyke Historien, Wetenschappen, Konsten en Handwerken* [Cabinet of nature and art, or cabinet of natural history, sciences, arts, and handcrafts]. Amsterdam 1719–1723 Vols. 1–8.

2) *Kern der Wissenschaften und brauchbarer Kunststücke* [Kernel of the sciences and useful works of art] Erfurt 1745. 2 vols. 8.

3) *Sammlungen nützlicher Nachrichten zur Beförderung der Künste und Manufakturen* [Collections of useful reports for the advancement of the arts and manufactures]. Jena 1755–1756. 8.

4) *Secrets des arts et metiers* [Secrets of the arts and trades]. Bruxelles 1755. Vols. 1 and 2. 12.
5) Johann Georg Model, *Chymische Nebenstunden* [Chemical leisure hours]. Petersburg 1762. Continuation in 1768.
6) Johann Georg Model, *Kleine Schriften* [Minor writings]. Petersburg 1773.
7) *Gemeinnütziges Natur- und Kunstmagazin, oder Abhandlungen zur Beförderung der Naturkunde, Kunst, Manufaktur und Fabriken* [Public magazine of nature and art, or treatises for the advancement of the study of nature, art, manufacture, and fabricks]. Berlin 1763–1767. Vols. 1–3.
8) William Lewis. *Zusammenhang der Künste, Gewerbe und Manufakturen, philosophisch-praktisch abgehandelt* [System of the arts, industry and manufactures, treated philosophically and practically], translated and with supplements by Johann Heinrich Ziegler. Zürich 1764–1766. 2 vols. 8.
9) *Neu eröffnete goldene Kunst-Pforte, oder allgemeines Magazin für Künstler, Handwerker, u.* [Newly opened golden portal of the arts, or general magazine for artisans, handcraftsmen, and others]. Nuremberg 1776. 2 vols. 8.
10) Johann Christian Polycarp Erxleben, *Physikalisch-chemische Abhandlungen* [Physical-chemical treatises]. Leipzig 1776. Vol. 1. 8.
11) Johann Christian Wiegleb. *Die natürliche Magie aus allerhand belustigenden und nützlichen Kunststücken bestehend* [The natural magic, comprised of various amusing and useful works of art]. Berlin und Stettin 1779. 8.
12) Franz Carl Achard *Chemisch-physische Schriften* [Chemical-physical writings]. Berlin 1780. 8.
13) *Die wohleingerichtete Kunst- und Werkschule, oder Handbuch für Künstler in fast allen nur erdenklichen Fällen* [The well-arranged school of art and handcrafts, or handbook for artisans in almost all imaginable cases]. 2 vols. 5th edition, Nuremberg 1784.
14) Friedrich August Cartheuser *Wahrnehmungen zum Nutzen verschiedener Künste und Fabriken* [Observations for the use of various arts and fabricks]. Gießen 1785.
15) Jakob Andreas Weber, *Physikalisch-chemisches Magazin für Aerzte, Chemisten und Künstler* [Physical-chemical magazine for doctors, chemists, and artisans]. Berlin 1780. Part one. 8.
16) Jakob Andreas Weber, *Bekannte und unbekannte Fabriken und Künste aus eigener Erfahrung* [Known and unknown fabricks and arts from personal experience]. Tübingen 1781. 8.
17) *Journal polytype des sciences et des arts* [Polytype journal of the sciences and the arts] Paris 1786.

18) Lorenz Florenz Friedrich von Crell, *Chemisches Journal für die Freunde der Naturlehre, Arzneygelahrtheit, Haushaltungskunde und Manufakturen* [Chemical journal for the friends of the study of nature, medicine, household management, and manufactures]. Lemgo 1778–1781. Parts 1–6.
19) Lorenz Florenz Friedrich von Crell, *Neueste Entdeckungen in der Chemie* [Most recent discoveries in chemistry]. Leipzig 1781–1784. Parts 1–12. 8.
20) Lorenz Florenz Friedrich von Crell, *Chemisches Archiv* [Chemical archive]. Leipzig 1784. Vols. 1 and 2. 8.
21) Lorenz Florenz Friedrich von Crell *neues chemisches Archiv* [Recent chemical archive]. Leipzig 1784–1785. Vols. 1–3.
22) Lorenz Florenz Friedrich von Crell *chemische Annalen* [Chemical annals]. Heimstädt und Leipzig. 1784. Vols. 1 and 2.
23) Lorenz Florenz Friedrich von Crell *Beyträge zur Erweiterung der Chemie oder zu den chemischen Annalen* [Contributions to the development of chemistry or to the chemical annals]. Heimstädt und Leipzig 1785.
24) Johann Samuel Halle *Magie, oder die Zauberkräfte der Natur, so auf den Nutzen und die Belustigung angewandt worden* [Magic, or the magical forces of nature, applied for use and amusement]. Berlin 1785–1786. 4 vols. 8.
25) *Technologisches Taschenbuch für Künstler, Fabrikanten und Metallurgen, aufs Jahr 1785 und 1786* [Technological pocket-book for artisans, fabricants, and metallurgists]. Göttingen 1785–1786. 8.
26) *Vollständige theoretische und praktische Geschichte der Erfindungen oder Gedanken über die Gegenstände aller drey Naturreiche* [Complete theoretical and practical history of inventions or thoughts about the objects of all three realms of nature]. Zürich. Part 1, 1786. Part 2, 1787. 8.

V Historical-Statistical Works

§73

1) *Reisen eines Philosophen, mit Betrachtungen über die Gebräuche, Künste, Ackerbau u. s. w. verschiedener Völker in Asien, Afrika*. [Travels of a philosopher, with observations over the customs, arts, agriculture, etc., of various peoples in Asia and Africa]. From the English. Danzig 1773. 8. (First published in French by Pierre Poivre as *Voyages d'un Philosophe*. Year 5.
2) Gabriel Jars (elder), *Voyages metallurgiques, ou recherches et observations sur les mines d'or et d'argent, celles de plomb, de cuivre, de bismuth, de cobalt et de mercure; les fabriques d'azur, de céruse, du blanc de plomb et du minium, faites depuis 1757 jusques et compris 1769, en Allemagne, Suede, Norwege, Hongrie, Tirol, Angleterre, Écosse, dans le Hartz, la Saxe, le Comté*

de Mansfeld, la Bohême et la Hollande [Metallurgical voyages, or investigations and observations on gold and silvermines, those of lead, copper, bismuth, cobalt, and quicksilver; fabricks of azure, ceruse (céruse, white lead), lead white (blanc de plomb), and minium (red lead), made since 1757, up to and including 1769, in Germany, Sweden, Norway, Hungary, Tirol, Enlgland, Scotland, the Harz, Saxony, Mansfeld (present day Saxony-Anhalt), Bohemia, and Holland. Lyon]. 1774–1781. Vols. 1–3. 4. German translation, *Metallurgische Reisen*, by Carl Abraham Gerhard. 2 vols. 1777.

3) Benedikt Franz Hermann. *Reisen durch Österreich, Steyermark, Kärnthen, Krain, Italien, Tyrol, Salzburg, und Baiern, im Jahre 1780* [Travels through Austria, the Steyermark, Karnten, Crain, Italy, Tirol, Salzburg, and Bavaria in the year 1780]. Vienna 1781. 8. Vols. 1 and 2.

4) Johan Jacob Ferber, *Neue Beyträge zur Mineralgeschichte verschiedener Länder* [New contributions ot the mineral history of various lands]. Mietau 1778. Vol. 1. 8.

5) August Friedrich Wilhelm Crome, *Europens Produkte, zum Gebrauch der neuen Produkten-Karte von Europa* [Europe's products, for the use of the recent product-map of Europe]. Dessau 1782. 8.

6) Johann Jacob Volkmann. *Neueste Reisen durch England, vorzüglich in Absicht auf Kunstsammlungen, Oekonomie, Manufakturen und Landsitze der Großen. Aus den besten neuesten Schriften zusammengetragen* [Most recent travels through England, primarly with regard to art collections, economy, manufactures and country manors of the wealthy. Collected from the best recent publications]. Leipzig 1781–1782. 2 vols.

7) Johann Jacob Volkmann, *Neueste Reisen durch Spanien, vorzüglich in Ansehung der Künste, Handlung, Oekonomie und Manufakturen* [Most recent travels through Spain, primarily with regard to the arts, trade, economy and manufactures]. Leipzig 1785. 2 vols.

8) Johann Jacob Volkmann. *Neueste Reisen durch die verinigten der Niederlande, vorzüglich in Absicht auf die Kunstsammlungen, Naturgeschichte, Oekonomie, und Manufakturen* [Most recent journeys through the united Netherlands, primarily with regard to the art collections, natural history, economy, and manufactures]. Leipzig 1783.

9) Benedikt Franz Hermann, *Abriß der physikalischen Beschaffenheit der östreichischen Staaten und des gegenwärtigen Zustandes der Landwirthschaft, Gewerbe, Manufakturen, Fabriken und der Handlung in denselben* [Outline of the physical composition of the Austrian states and the current condition of their agriculture, industry, manufactures, fabricks, and trade]. Petersburg und Leipzig 1782. 8.

10) *Materialien zur Geschichte des Vaterlandes, dessen heutige Geographie, Nahrungsstand, alte Sitten und Gebräuche in verschiedenen Gegenden Baierns, dann der Herzogthümer obsern Pfalz, Neuburg und Sulzbach* [Materials on the history of the fatherland, its current geography, income, old mores and customs in various regions of Bavaria, and then the principalities of upper Pfalz, Neuburg, and Sulzbach]. Published by the office of intelligence of Munich since 1782. 4.
11) Friedrich Nicolai, *Beschreibung der Königlichen Residenzstädte Berlin and Potsdam* [Description of the royal seats of Berlin and Potsdam]. 3 vols. Third edition. Berlin 1786. 8.
12) Friedrich Nicolai, *Beschreibung einer Reise durch Deutschland* [Description of a voyage through Germany]. Berlin und Stettin, 1784–1786. 6 vols. 8.
13) Paul von Stetten (the younger), *Kunst- Gewerb- und Handwerksgescihcte der Reichsstadt Augsburg*. Augsburg [History of the arts, industry, and handcrafts of the imperial city Augsburg]. 1779. 8.
14) Leonhard Meister. *Beyträge zur Geschichte der Künste und Gewerbe, der Sitten und Gebräuche* [Contributions to the history of the arts and industry, mores, and customs]. Zürich 1774. 8.
15) Johann Beckmann, *Beyträge zur Geschichte der Erfindungen* [Contributions to the history of inventions]. Leipzig 1780–1786. 6 vols.
16) Christoph Gottlieb von Murr. *Journal zur Kunstgeschichte und allgemeinen Litteratur* [Journal for the history of the arts and general literature]. Nuremberg since 1775. 15 vols. 8.
17) August Friedrich Wilhelm Crome, *Abhandlungen aus dem Handlungsgebiet zur Kenntniß und Geschichte des Fabrik- und Kommerzwesens* [Treatises from the area of trade for the knowledge and history of fabrick and commercial matters]. Leipzig 1786.

PART 3

General vs. Historical Technology, circa 1800

ESSAY

Technology's Two Perspectives: the General and the Historical

The two translations in this section are of texts published in the first decade of the nineteenth century. Together, they exemplify the two perspectives that dominated technological writing at that time: one concerned with a general theory of technology, the other interested in situating the concept of technology within historical narratives of progress. It is tempting to connect this particular moment in the conceptual history of technology to later events in the nineteenth century, but in this case the perspective of hindsight can also have an obfuscating effect. From the present vantagepoint, the nineteenth century is crowded with innumerable technological advancements in production and theory. To the former category belong the perfection of the steam engine, the widespread mechanization of industry, and advancements in communication technologies, among others. To the latter category belong the significant milestones in philosophical reflections on the interrelations between technology, the human body, and society, including Marx's *Kapital* (1867), and Ernst Kapp's groundbreaking *Grundlinien einer philosophie der Technik* [Elements of a philosophy of technology] (1877), to give just two prominent examples. In the first decade of the nineteenth century, however, the situation on the ground looked quite a bit different. It was shaped by questions of how to organize and distinguish technological objects, even as the idea of the technological object is arguably becoming broader, as well as by questions directed towards the various strengths and weaknesses of the technological systems being proposed. As John Durham Peters has argued, technological determinism doesn't have to be a dirty word, and the emphasis of this project is to take this writing on its own terms rather than argue how late-eighteenth and early-nineteenth century writing on technology either anticipates or is genealogically linked to philosophical developments of the late nineteenth century.

The first translation is of Johann Beckmann's "Draft of a General Technology,"[1] which was published in 1806, almost thirty years after the *Guide to*

1 In some cases, to translate is to choose between several possible terms, each with its own merits and weaknesses. Such is the case with one of the most significant choices of this project: how to translate the word *Entwurf* Johann Beckmann uses in his essay on a general technology. "Project" – which is also the word used in the recently-published French translation

 | DOI:10.1163/9789004549548_012

Technology.[2] It is the first published English translation, but well behind work on the "General Technology" already accomplished in France, where a translation accompanied by several essays was published in 2017 in a volume co-edited by Guillaume Carnino, Liliane Hilaire-Pérez, and Jochen Hoock.[3] In broad terms, the aim of that volume is to reflect on the historical concept of technology as "science of action and activity" – or "human science" – while underscoring is both the "originality" and "opportunism" of Beckmann's project, with its links to "the production of technical knowledge by administrations" and "efficacious management of resources of the state."[4] More specifically, the editors of that publication would like to show, as Jochen Hoock has argued, that the general technology is more than a simple reworking of the *Guide to Technology*, as

---

of the general technology – is a safe way to go, and allows for a parallel between the *werfen* and *jacere*, which both mean "to throw." To this translator's ears, however, "draft" is a better choice. When one consults the entry on *Entwurf* in Adelung's German dictionary, one finds the following definitions and examples: a "depiction, ordering of the essential parts of a future whole" (*Abbildung, Anordnung der wesentlichen Theile eines künftigen Ganzen*), such as one would make "of a painting, a building, a sermon, a poem, [or] a letter" (*eines Gemäldes, eines Gebäudes, einer Predigt, eines Gedichtes, eines Briefes*). It is also the first composition of a piece of writing (*der erste Aufsatz einer Schrift*). One will note that "project" works well at first as a translation, when one is speaking in terms of abstractions (like the parts of a future whole), but "draft" fits better once the particular examples are given: a draft of a poem, a sermon, etc. A final reason for choosing "draft" over "project" or even "'outline" as a translation for *Entwurf* lies in Beckmann's own understanding of what a general technology is: a *Verzeichnis*, usually translated as "index" or "catalog," and which contains the root verb *zeichnen*, to draw or sketch. For Ebers, the act of recording, or *aufzeichnen*, encompasses the meanings of taking inventory, sketching, and making a "rough Draught." The choice to translate the *Entwurf einer algemeinen Technologie* as the "draft of a general technology" is intended to preserve the forward-looking sensibility of a "project," while at the same time keeping in mind that at heart, this technology is a written record that attempts to maintain a balance between very different things: abstract reflections on systematic order on the one hand, and focused discussions of those tools and those objects which are fundamental to technical practices on the other.

2 He had not been idle in the interim years, in which he published writing on the history of inventions and the science of trade, including the following titles: *Beyträge zur Geschichte der Erfindungen*. 5 vols., Leipzig/Göttingen 1780–1805; *Anleitung zur Handlungswissenschaft* (Guide to the science of trade) Göttingen 1789; *Physikalisch-oeconomische Bibliothek* (Physical-economical library) 23 vols., Göttingen 1770–1807; *Beiträge zur Oeconomie, Technologie, Polizey- und Cameralwissenschaft* (Contributions to economy, technology, governance and cameralist science), 12 vols., 1777–1791.

3 In the introduction to their volume, Carnino and Hilaire-Pérez describe their initiative to reconsider the "forgotten meaning of technology" as a "science of technics" (*science de la technique*) or a "science of the arts" (*science de les arts*) (introduction to *La Technologie générale*, 13).

4 Carnino and Hilaire-Pérez, introduction to *La Technologie générale*, 14.

previous scholars have argued. In other words, it is "no longer just a matter of a comparative systematization of different effective technical processes and of their transfer in the context of a precise operation, but of the attempt to liberate from them the principles for the functioning and the transfer to other domains of industrial activity on a much larger scale."[5]

Beckmann did not coin the phrase "general technology" (*allgemeine Technologie*) in German. Georg Lamprecht already defines general technology in his 1787 *Instruction Manual to Technology* using expansive terms: it encompasses "the theories of all kinds of handcrafts … that exist in the world, without distinction."[6] Other technological writers also took up the term, including the German botanist and cameralist Friedrich Ludwig Walther (1759–1824) and Daniel Heinrich Bensen (1761–1805), who was a professor of cameralist science, philosophy, and law at Erlangen. Walther, writing in 1796, defines "general technology" as follows:

> This general part of technology makes us familiar with different kinds of artisanal trades (*Kunstgewerbe*), with the basis of the artisanal economy, with its useful publications, with the required persons, buildings, and tools for the encouragement of the artisanal economy, and with the processes of managing it financially.[7]

Whereas Walther's idea of a "general technology" essentially requires an expansive economic how-to manual geared towards that area of industry occupied by handcrafts, fabricks, and manufactures of various kinds, Bensen, writing in 1799, offers a somewhat different perspective:

> With regard to the purpose, which can be different in the scientific form of technology, as the totality (*Inbegriff*) of the most varied knowledge deriving from experience, the general technology must be separated from the particular. All rules and principles (*Grundsätze*), therefore, which are adopted and followed by all or at least by most handcrafts for the most agreeable conducting of their trade, are to be regarded as an independently standing whole, and to be set up as such.[8]

5 Hoock, "Johann Beckmann et la place du discours technologique," 109.

6 P. 120.

7 Friedrich Ludwig Walther, *Versuch eines Systems der Cameral-Wissenschaften* (Attempt at a system of the cameralist sciences), vol. 3 Technologie (Georg Friedrich Heyer, 1796), 11.

8 D. Heinrich Bensen, *Versuch eines systematischen Grundrisses der reinen und angewandten Staatslehre für Cameralisten* (Attempt at a systematic outline of the pure and applied political theory), vol. 1 (Johann Jakob Palm, 1798), 205.

Unlike Walther, Bensen is more concerned with a theoretical framework that encompasses a constellation of related concepts, including scientific form, experience-based knowledge, and a set of rules and principles broad enough to be applicable to most kinds of handcrafts. Beckmann's own undertaking navigates between both points of view by acknowledging both the "parts" and the "whole" of a general technology. He does this by integrating the idea of a far-reaching technological gaze with a concern for how to organize technological objects: "Whoever has made a study of learning many handworks and arts," he writes, "and whoever has trained himself to survey many with a single glance, must notice that many handcrafts, however distinct their materials and wares may be, nonetheless have to carry out several works for a common purpose. Or, that they know how to reach a common purpose in very different ways."[9] With regard to the order of the parts of technology, he proposes the idea of an inventory (*Verzeichniß*):[10] "of all the various purposes that the handcraftsmen and artisans have in their various works, and simultaneously an inventory of all the means through which they know how to attain each of them."[11]

The inventory Beckmann proposes *is* the general technology. It makes no claims to completeness, yet provides enough detail for readers to understand his method. Beckmann takes as his example the category of "making things smaller" (*zerkleinern*), before defining subcategories based on the quality of physical body in question (brittle, soft, fibrous, etc.), and then providing lists of the particular processes that would be appropriate for that kind of material.[12]

9 P. 166.

10 Contemporary German usually understands a *Verzeichnis* to be an index, list, table, or similar typographical object, but the eighteenth-century understanding is somewhat different. Ebers' dictionary translates the noun, inventory, as a "Verzeichniß beweglicher Güter" and the verb, to inventory, as "in ein Verzeichniß bringen." Ebers, vol. 3 (1799), s.v. "Verzeichniß."

11 P. 166.

12 Danielle Candel and John Humbley have emphasized the way in which Beckmann's project is related in various ways to dictionaries and other kinds of reference works of his time, arguing that "Even if the *Entwurf* was not conceived as a dictionary, it includes numerous functions of them, notably definitions, explications of words, equivalents from other languages, as well as, more generally, a reflection on questions of denomination." Danielle Candel and John Humbley, "L'émergence d'une conscience terminologique: regard métalexicographique porté sur l'Entwurf der algemeinen Technologie de Johann Beckmann (1806)" [The emergence of a terminological conscience: a meta-lexicographic look at the draft of a general technology by Johann Beckmann], in *La technologie Générale*, ed. Guillaume Carnino, Liliane Hilaire-Pérez, and Jochen Hoock, trans. Joost Mertens.

For example, under the category "making smaller" and the sub-category "brittle bodies," Beckmann includes a number of relevant terms:

- Zerkleinern (*making smaller*)
  - Spröde Körper (*brittle bodies*)
    - Zerschlagen (*striking into pieces*)
    - Zerhauen (*chopping*)
    - Zerbrechen, Zerbröckeln (*shattering, crumbling*)
    - Zerdrücken (*crushing by pressing into pieces*)
    - Zertreten (*stamping*)
    - Zerwalzen (*milling*)
    - Zerstoßen (*crushing by pounding*)
    - Zerreiben (*grinding*)
    - Zerraspeln, Zerfeilen (*chipping, filing*)
    - Zermahlen (*grinding*)
    - Zersägen (*sawing*)
    - Zerhobeln (*planing*)
    - Zerdrehen (*separating by turning*)
    - Zerbeissen, Zerkauen (*biting, chewing*)
    - Zerspalten, Reissen (*splitting, tearing*)

One can readily see that Beckmann's general technology prioritizes processes over objects (in other words: technological verbs, as opposed to nouns).[13] It is also noteworthy that the examples he chooses contain the German prefix *zer-*. Adelung's *Historical-Critical Dictionary* defines this prefix as one which designates "a separation" and a "dissolution of the parts through the concept of the verb."[14] The *zer-* prefix, in verb constructions, embodies the technological counterpoint to the analytical process.

If one takes a broader point of view, it is an interesting exercise to compare Beckmann's categorization of techniques for breaking things apart with analogous moments in other discursive contexts. As an example, one could recall the phrase *disjecta membra poetae* borrowed from Horace's *Satires* – a work

---

Introduction by Guillaume Carnino and Liliane Hilaire-Pérez (Presses Universitaires, 2017), 218.

13 In a more general sense, this idea connects to "technical knowledge," defined by Liliane Hilaire-Pérez: "a science of the project," a matter of "functionality and adaptation" (Hilaire-Pérez, "Technology as a Public Culture in the Eighteenth Century," 136).

14 *Grammatisch-Kritisches Wörterbuch der Hochdeutschen Mundart, Ausgabe letzter Hand* (Final edition), Leipzig 1793–1801), digitalized version in the dictionary network of the Trier Center for Digital Humanities, Version 01/23, https://www.woerterbuchnetz.de/Adelung (accessed August 18, 2023), s. v. "zer-."

Beckmann also quotes from in the same volume where the *Draft of a General Technology* was published.[15] In many cases, it is used in contexts which display an aesthetic interest in the fragmentary, or to connect a literary idea to scientific method, as Johann Wilhelm Ritter does when he wishes for his time to organize his own scattered observations into a statement of "basic organic law."[16] In Ritter's case, the scientific notes and fragments themselves, the snippets of text, are the objects of interest that can be combined into a theoretically powerful whole. It's a useful analogy to the category Beckmann creates of physical/analytical processes, where instead of individual pieces he defines the processes themselves: a technological "object" need not be a corporeal one.

The purpose of the general technology is not just to list and categorize, but also to teach "in however many ways, and with however many tools," Beckmann writes, "how bodies of various kinds are smoothed, carded, shredded, netted, dried, straightened, bent, hardened, tempered, stiffened, thickened, loosened, thinned, sieved, warmed, or cooled."[17] A notable absence in the "Draft of a General Technology" is any direct mention of Carl Linnaeus, whose work was front and center in the *Guide to Technology*.[18] Tangibly present, both in name and through the concept of "translation," is Francis Bacon. In what is essentially a paraphrase from the *New Organon* (and his own reception of it in the *Guide to Technology*), Beckmann writes that the "translation of means (i.e., technological processes) from one handcraft to another" is just as important as new inventions.[19] Acts of translation such as Beckmann describes are designed to facilitate the requisite bird's-view, the gaze of a generalist. Even if they are first proposed by a scholar, rather than an artisan, the aim is to allow master craftsmen to integrate the practice of translation into their particular handcrafts, although, as Beckmann acknowledges, "this would occur most

15 *Vorrath Kleiner Anmerkungen über mancherley gelehrte Gegenstände* (Collection of small observations concerning various scholarly topics), vol. 3 (Friedrich Röwer, 1806), 410.

16 In the preface to his fragment collection, Johann Wilhelm Ritter (referring to himself), describes a desire to go through the many notes and fragments written by the physicist, to "collect the *Disjecti membra Poëtae*" and "to order them according to that basic organic law which no book teaches yet." *Key Texts of Johann Wilhelm Ritter (1776–1810) on the Science and Art of Nature*, trans. Jocelyn Holland (Brill, 2010), 113.

17 Beckmann, "General Technology," p. 166.

18 Hans-Peter Müller finds ample evidence of Linnaeus's presence in the general technology, even if Linnaeus is not mentioned by name, saying that it contains "a binomial technological systematics, completely in Linné's sense of it, arranged according to genuses and species of purposes and means." Hans-Peter Müller, "Beckmanns Rekombinatorik und die Ars Inventiva des Raimundus Lullus" (Beckmann's recombinatorics and the art of invention of Raimundus Lullus). Hans-Peter Müller, *Johann Beckmann (1739–1811)*, 105.

19 Pp. 167–8.

infrequently until more handcrafts, through the study of auxiliary sciences and general technology, have been formed into general concepts."[20]

As he does in his earlier work, Beckmann draws from a range of sources in his "Draft of a General Technology." There is a nod to the classics, such as Seneca's *Epistles*, Plato's *Republic*, and others, with the purpose of using the authority of classical sources to elevate the work of the craftsmen. From Seneca's *Epistles*, Beckmann applies a scientific narrative – that "the application of scientific study and classification of the discoveries made by others" can constantly generate new ideas – to the technological context, also quoting directly Seneca's claim that the inventiveness of craftsmen is worthy of respect. And from Plato, Beckmann borrows the idea that the world will be a better place when kings become philosophers and philosophers kings to craftsmen and scholars. As difficult to implement as those ideas might have been from a practical point of view, they indicate a kind of social idealism without, however, directly addressing the notion of a "general good" – a phrase not used by Beckmann himself, but which one can find in both Lamprecht's and Poppe's work.

Beckmann's perspective on the discourse of medicine can also provide some insight into the organization of the "General Technology." From a historical perspective, the relationship of medicine to technology has been a contested one. In the *Guide to Technology*, Beckmann briefly mentioned "apothecary art" (*Apothekerkunst*) within an untitled category that included the preparation of wines, beer and vinegar brewing, the preparation of liqueurs, and the like. In §12 of the *Guide*, however, Beckmann states that "a principal point of technology is the correct determination of main and secondary materials" which he refers to collectively as the "*materia technologica* or the study of materials," before comparing them – unfavorably – to the *materia medica*, for being less complete.[21] Some technologists, such as Lamprecht, exclude the apothecary arts in their entirety from technology, but Beckmann still finds the term useful – to a degree. Writing from a later historical perspective, Beckmann suggests that the general technology he proposes seems "to have great similarity with the systematic catalogues of remedies, the so-called *materia medica*."[22] He defines those *materia medica* which contain the *simplicia*, i.e., means of producing single effects (such as sweating, purging, etc.). These are presumably comparable to single verbs in general technology, for individual works (*Arbeiten*). At the same time, Beckmann distances himself from the name *materia*

20 P. 168.

21 P. 170.

22 P. 169.

technologica ("whose name" he does "not wish to recommend").[23] One speculation as to why the term *materia technologica* is not recommendable has to do with the question of what, precisely, constitutes the "materia." The conventional eighteenth-century definition of the *materia medica* focused on physical substances: bodies that belonged "partly to the *fossile*, partly to the vegetable, and partly to the *animal Kingdom*," as was mentioned above in the introductory essay to Part One.[24] As has been argued above, however, Beckmann's technology is very much process-oriented, and therefore organized in terms of techniques first before considering the physical objects that undergo refinement. Whereas, from the point of view of the *Guide to Technology*, the *materia technologica* is envisioned as somewhat less complete than the *materia medica*, from the point of view of the general technology one could say that technology's *materia* needs to be defined more abstractly than its medical counterpart.

Each of the technological writings discussed in this volume at some point laments the fact of its own incompleteness, and Beckmann's "general technology" is no exception: "The works of the collective handcrafts and arts, and the tools and machines belonging to them, are much too numerous to allow anyone to believe that he could be able to know and classify them all."[25] As Sean Franzel, Ilinca Iurascu, and Petra McGillen have shown in a project on nineteenth-century inventories, this sense of incompleteness can be understood in terms of a genre convention: "Inventories," they argue, "promise an overview and a systematic organization of accumulations of discrete objects, yet are based in a sense of incompleteness vis-à-vis the excess of practices of storing, listing, ordering, and reordering with which they are confronted."[26] This observation applies just as well to inventory items that are physical objects in historical collections as it does to more abstract inventories of the technological manuals around 1800.

1 Poppe's *History of Technology*

The final translation of this volume comes from Johann Heinrich Moritz von Poppe's *History of Technology* (1807).[27] Poppe's status in the history of

23 P. 170.

24 Hill, *A History of the Materia Medica*, 1.

25 P. 173.

26 Sean Franzel, Ilinca Iurascu, and Petra McGillen, "Media Inventories of the Nineteenth Century: A Report from Two Workshops," in *Goethe Yearbook* XXVIII (2021): 285.

27 Johann Heinrich Moritz Poppe, *Geschichte der Technologie seit der Wiederherstellung der Wissenschaften bis an das Ende des achtzehnten Jahrhunderts* (History of technology since

technology has been a matter of some debate since the nineteenth century. The German jurist Robert von Mohl's harsh criticism that "Poppe, the technologist, was scientifically a zero" has done its part to taint his reception.[28] One can find certainly appreciation for Poppe's historical work, such as the reference to the "admirable mass of data" it compiles, as Karl Karmarsch noted in his *History of Technology*; at the same time, such faint praise often gives way to more critical commentary, as Helmut Kahlert has observed in an essay on Poppe aptly subtitled an "appreciation of a disputed personality."[29] In the past decades, however, there has been an attempt to take a more nuanced look at Poppe's contributions to the science of technology at the beginning of the nineteenth century. Rolf Seubert has speculated that the harsh judgements about Poppe's work "personalize a dilemma about the scientific approach to technology"[30] and that such judgements against Poppe "scarcely take into account the objective stage of development of handcraft and industry and not at all the stage of development of technology as a science."[31] The goal of the present volume is to show the value of a history of technology itself at the beginning of the nineteenth century, and how a historical perspective may be productively balanced alongside theoretical reflections on technology.

Born in 1776, the son of a watchmaker who was also a university mechanic in Göttingen, Poppe's early educational environment naturally transcended any divisions between the workshop and the schoolroom. He continued to assist his father even after beginning his studies, and his first publication, in 1797, was devoted to the art of watchmaking. Works on various other topics followed,[32]

---

the restoration of the sciences through the end of the eighteenth century), vol. 1 (Johann Friedrich Röwer, 1807).

28 Robert von Mohl, *Lebenserinnerungen* (Life remembrances), vol. 1 (Deutsche Verlags-Anstalt, 1902), 166.

29 Helmut Kahlert. "Johann Heinrich Moritz von Poppe (1776–1854). Würdigung einer umstrittenen Persönlichkeit" (Evaluation of a disputed personality), in *Zeitschrift für Würtetembergische Landesgeschichte* (Periodical of Würtemberg's provincial history), ed. Hans-Martin Maurer, 64 (W. Kohlhammer, 2005), 211–226. Kahlert's essay contains both a throrough overview of Poppe's literary output and a biography of Poppe's key contributions to German academic culture, along with a bibliography of recent publications that reassess Poppe's work.

30 Rolf Seubert. "Handwerkbildung als Beitrag zum Aufbau einer Nationalindustrie. Johann Heinrich Moritz Poppe und der Zusammenhang von Technologieentwicklung und beruflicher Bildung" (Handcraft education as contribution to the building of a national industry), in *Johann Beckmann (1739–1811)*, ed. Günter Bayerl und Jürgen Beckmann (Waxmann, 1999), 187.

31 Seubert, "Handwerkerbildung," 188.

32 In addition to a second, revised history of the art of watchmaking, Poppe also published the *Allgemeines Rettungsbuch* (General rescue book, 1805), which is considered one of

leading up to the *Handbuch der Technologie* (Handbook of technology), whose first volume was published in 1806. This work was intended for instruction in the gymnasium or university, and its language adheres closely to Beckmann's own script (see esp. §31). In 1807, when Poppe finished the *History of Technology*, he had already read Beckmann's general technology, published the year before. The part of Poppe's *History of Technology* selected for translation is the first section of this work, which Poppe designated as a "general introduction to technology." In addition to showing how reflections on technology can be closely integrated with historical perspectives, the moral questions Poppe raises in his history will also help readers make connections with the questions of the moral values of *technê* that have informed the historical usage of this term.

It is important to note that Poppe begins his history not with a reference to the Greeks and Romans, or to other civilizations known for their technical expertise, but with a fictional narrative reminiscent of Rousseau's *Discourse on Inequality*. Just as Rousseau speculated that a "savage who is robust, agile, courageous," and "armed with stones and a good stick" could have proven to ferocious beasts that he is "just as ferocious as themselves,"[33] Poppe begins with a narrative where the original humans, on the "as of yet uncultivated orb," could nonetheless find any necessary means of protection: "Against wind and weather they would know how to look for a shelter," he muses, and "at most, wild animals would cause them discomfort, but soon they would stand as conquerors against these as well."[34]

In this philosophical fiction, the idea of tool use is more important for Poppe, the technologist, than for Rousseau, for whom human limbs weaken when tools are used to supplement them. The second paragraph of Poppe's "general introduction" integrates technology indirectly into this historical fiction, through the idea early humans would have had to "*further refine products of the earth* (*Naturalien*) in order to satisfy their needs in a more complete way" (§2, i.i.o.), along with the idea that the role of human spirit is to allow man to reflect on what is lacking and, through the spirit of invention, to devise a way of supplementing this lack.[35] What follows is a historical narrative equally invested in a "history of the spirit" and the particular cultural contexts it inhabits – moments in human history that at times work in

---

the first publications relating to workplace injuries, and the *Encycklopädie des gesamten Maschinenwesens* (Encylopedia of all types of machinery, 1803f.).

33 Rousseau. "Discourse on the Origins or Foundations of Inequality among Men," in *The Major Political Writings of Jean-Jacques Rousseau*, trans. and ed. John T. Scott (University of Chicago Press, 2012), 67.

34 P. 200.

35 P. 200.

harmony with the spirit of invention, and at times do not. Poppe's unapologetically German-centric perspective focuses on the emergence of guilds, with an emphasis on the German states' special status at the time (given that they had, according to Poppe, "the greatest masters in almost every handcraft" and were responsible for "most inventions").[36]

A quick note on the social context – artisans and guild status – will help give some social-historical contour to Poppe's perspective. As Liliane Hilaire-Pérez has observed, "one main feature of the Industrial Enlightenment was the active part played by artisans in the process of technical intensification. They did not form a homogeneous group, either culturally or economically. Their culture was as complex and hybrid as that of the engineer or industrialist. They were involved in open technique patterns, both institutional (with guilds acting sometimes positively) and commercial."[37] In addition, not only artisans, but also guilds themselves have a contested history. Over time, there have been numerous vocal critics. In the eighteenth century, Adam Smith asserted in his influential book, *The Wealth of Nations*, that craft guilds are a "conspiracy against the public," and guilds were abolished altogether in France in the course of the French Revolution.[38] In the historiography of the twentieth century, guilds have also found vocal critics such as Joel Mokyr, who blamed guilds, "together with tax collectors and foreign invaders, for the industrial decline of Northern Italy, Southern Germany, and the Low Countries."[39] In their study, *Guilds, Innovation and the European Economy, 1400–1800*, S. R. Epstein and Maarten Prak show critics of Mokyr's argument have suggested that guilds were in fact flexible enough to adapt to changing environments and go a step forward with a more rigorous economic argument that claims that "the impact of guilds on the early modern economy was more positive than has so far been acknowledged by historians of the traditional, and even of the revisionist, school."[40] Sheilagh Ogilvie, while acknowledging the differing perspectives in recent scholarship, sums it up more laconically: "guilds tended to do what was best for guild members."[41]

36 P. 206.

37 Hilaire-Pérez 2007, 146.

38 S. R. Epstein and Maarten Prak, "Introduction," *Guilds, Innovation and the European Economy, 1400–1800*, ed. S. R. Epstein and Maarten Prak (Cambridge University Press, 2008), 1.

39 Epstein and Prak, "Introduction," 2.

40 Epstein and Prak, "Introduction," 23.

41 This could include, among other things: "reducing threats from innovators, competitors, and audacious upstarts" and "generating sufficient rents to pay off the political elites that enforced guilds' privileges and might otherwise have interfered." Sheilagh Ogilvie, *The European Giulds. An Economic Analysis*. (Princeton University Press, 2019), 17.

In Poppe's narrative, Nuremberg receives special mention for its diversity of crafts and far-reaching export economy. At the same time, Poppe does not shy away from structural points of weakness in the guild system, ranging from the excessive secrecy that impedes the transmission of ideas regionally and between crafts, to the cases of physical abuse endemic to guild rituals. When Poppe's narrative arrives at the eighteenth century, he uses a familiar image to help integrate theoretical reflections into a purely historical account: the idea of the scholar who explores the space of the workshop. And just as has been observed in the work of Beckmann and Lamprecht, for Poppe, too, the proposed exchange of ideas represents the ideal of scholars learning from craftsmen, just as craftsmen could benefit from mechanics, the history of nature, chemistry, and physics. The broadening from historical to more speculative thinking occurs in §42, when Poppe shifts from describing how the culture and practice of the arts have developed in Germany over time to a bird's-eye view that attempts to see centuries worth of technical knowledge as a whole: "A single science has emerged from the knowledge of handcrafts, manufactures, and fabricks."[42] With a nod to older notions of the history and the science of the arts (*Kunstgeschichte* and *Kunstwissenschaft*), Poppe concludes §42 with a succinct and impactful observation: "(Technology) is distinguished from art itself, just as knowing is from doing, or theory from praxis."[43] Poppe thereby signals his willingness to elevate technology's status by defining it in terms of general knowledge, rather than through discrete practices that could be reduced to a specific handcraft, manufacture, or fabrick.

A close look at Poppe's technical language also reveals a term, the *Handgriff*, that can be used as an example to support the idea that there is a theoretical dimension to technology that is connected to, yet distinct from, artistic practice. Already in the eighteenth century, this term encompassed both a physical definition, "handle," and metaphorical ones, where it was adapted to a wide range of semantic meanings, including "technique," "skill," "knack," and "trick" – meanings already associated with the concept of *technê*. Jacobsson's *Technological Dictionary* defines *Handgriff* as a kind of tacit knowledge in the sense akin to Richard Sennett's understanding of it, something which Jacobsson describes as useful for

> small skills, which are learned only through experience and practice, and which cannot always be described, whereby the craftsman and artisan completes his wares. As trivial as they may seem to be, still time and

42 P. 235.

43 P. 235.

> patience are required for acquiring and practicing these small skills. In many workshops, techniques (*Handgriffe*) are the most important thing.[44]

Poppe, whose use of *Handgriff* aligns with Jacobsson's definition, takes advantage of the term much more frequently than either Beckmann or Lamprecht. For Poppe, the *Handgriff*, as technique, is part of the secret knowledge of those craftsmen who "must swear an oath not to reveal any technique (*Handgriff*) or any invention to a foreigner" (§29); it is part of the essential quality of each handcraft (whereby Poppe lists "methods, proper techniques (*Handgriffe*), precautions, and judgements," §31); it is of interest to those scholars, especially in times past, who wanted to augment their knowledge of handcrafts (and "found it worth their trouble to think more about various kinds of work, techniques (*Handgriffe*), and advantages," §41); it connects technological practice with scientific theory (via the idea that "many skillful techniques (*Handgriffe*) and the completion and use of many tools rely on *mechanical theories* which can best be learned scientifically," §41); and finally, it is an important component of the objective technological gaze advocated by Poppe, one that will allow for a more economical approach to applying skills from one handcraft to another (i.e., through the idea that "many handcrafts, as different as their materials and wares are, accomplish various works for the same purposes, and require for that the same techniques (*Handgriffe*), the same instruments, and the same machines and other means," §45). In short, there is an argument to be made that initial theoretical reflections on technology in Poppe's thought take their point of departure from the metaphor of the *Handgriff*.

2 Conclusion

Just as Poppe's historical narrative begins ahistorically, with a vision of man in the state of nature, so too does it take a decidedly ahistorical turn at the end when it replaces its chronological account of human technological progress with a reflection on various systems of technological order that have been attempted in the years prior to the publication. Poppe's reflections on his own approach to technology are embedded within the project itself. In a gesture which should be familiar by now, given that Beckmann's and Lamprecht's

44 Jacobsson, p. 204. *Technologisches Wörterbuch oder alphabetische Erklärung aller nützlichen mechanischen Künste, Manufakturen, Fabriken und Handwerker.* Vol. 2, from G to L. (Friedrich Nicolai, 1782).

writings also testify to a dissatisfaction where systems of technology are concerned, Poppe also expresses his frustration as the as he attempts to get a view of the whole. Such dissatisfaction can be understood both in theoretical and practical terms as the dissatisfaction of a science for which any attempts at abstraction are hampered by the proliferation of the technological objects keeping it from taking flight. At the same time, such attempts are worthy of our attention because they signal a widespread phenomenon in the conceptual history of technology at the end of the eighteenth century.[45] The dissatisfaction expressed by the self-described technologists of that time also generated a considerable amount of intellectual energy that was invested in the idea of technology in various ways: as a new word in German, as a concept, as an academic discipline, and as an emerging science struggling with self-definition. Their *Technologie* is not ours, but its essential hybridity – as something rooted in the world of empirical objects while making claims to theoretical aspirations – should guarantee it a greater place than has previously been acknowledged in the history of technology.

45 In addition to lamenting his own limitations, for example, Poppe also refers to problems in the technological systems of Jung, Rosenthal, Walther, Bensen, the Rösling brothers and others, in addition to Beckmann and Lamprecht.

Johann Beckmann's *Draft of a General Technology*, 1st Edition (1806)

Technology teaches one to prepare both raw and already processed materials for all of the most varied kinds of use which people know to make of them.[1]

The preparation for a particular use is called the *processing* (*Verarbeitung*) of the material. Those who make a trade of it are called *handcraftsmen*, and such a trade a handcraft, to which also belong, in a scientific understanding, various arts, the fabricks, and manufactures.

Before a material becomes that which it is supposed to be, it requires various works. One tends to say that a sheet of paper has to go through the hands 30 times before it is completely prepared. I will not count again to see if this has been properly calculated. Rather, I want to name a few of the different kinds of work, in order to explain what I mean by the term. The sorted rags are chopped up through hacking on the block, through cutting in the rag-cutter, through stamping in the vessels, and through grinding in the paper mill. The finely divided mass, the pulp, is brought together again or collected into sheets, which one calls drawing up the pulp. The sheets must then be smoothed by pressing, striking, etc. One has there at least three different kinds of work.

Whoever has made a study of learning many handworks and arts, and whoever has trained himself to survey many with a single glance, must notice that many handcrafts, however distinct their materials and wares may be, nonetheless have to carry out several works for a common purpose. Or, that they know how to reach a common purpose in very different ways. As Geber has already commented: *multae sunt viae ad unum inventum et unum finem* (many are the ways to a one invention and one end).[2]

The planing of the carpenter, the grinding of the spectacle maker, the beating of the book binding, the pressing of the cloths, the glossing of the linen, as distinct as all of these procedures seem to be and really are, they still have

1 *Entwurf der algemeinen (sic) Technologie* [Draft of the general technology] in Beckmann, *Vorrath kleiner Anmerkungen über mancherley gelehrte Gegenstände* [Collection of small observations concerning various scholarly topics], vol. 3 (Johann Friedrich Röwer, 1806), 463–533.

2 *The Summa perfectionis of Pseudo-Geber: a critical edition, translation and study*, ed. and trans. William R. Newman (Brill, 1991), 667.

 | DOI:10.1163/9789004549548_013

a common purpose, namely, to make bodies smooth, although in a few cases more than one purpose will occasionally be reached, which I will however not attend to here.

At the moment I wish for an inventory (*Verzeichniß*) of all the various purposes that the handcraftsmen and artisans have in their various works, and simultaneously an inventory of all the means through which they know to attain each of them.

Such an inventory I would give the name of a *general technology*, or the *first or general part* of *technology*. The particular part would contain the description of individual handcrafts.

The former (general technology) would teach – in however many ways, and with however many tools – how bodies of various kinds are smoothed, carded, shredded, netted, dried, straightened, bent, hardened, tempered, stiffened, thickened, loosened, thinned, sieved, warmed, or cooled (1)[3] and how they are made more transparent and opaque, more elastic, more pliant, etc. Furthermore, it would teach the means by which fluid bodies are made clear, decolored, vaporized, and more malleable. With regard to the latter, for example, one would do this by using rubber, fleawort, quince seeds, fenugreek, clover, clay, egg white, sugar, etc.

No one who is knowledgeable of these objects will doubt the usefulness of such an inventory. It certainly seems that every handcraft has long since invented appropriate means for its works, and truly one has to admire the acumen and spirit of invention which reveals itself, especially since scholars have had no part at all in it. Seneca has said quite correctly: non ab aliis excogitata ista sunt, quam a quibus hodieque curantur.[4]

3 (Translator's note: all of Poppe's own footnotes begin with a numeral printed in parentheses). (1) Just as *erwärmen*, for making warm, is derived from *warm*, so too can one also from cold say *erkälten*, for making cold, or *abkühlen*. This occurs intentionally, sometimes quickly, sometimes slowly; for example: in the glass houses of the glassmaker, in the cooling tub and the moor's head (*Mohrenkopf*, a water cauldron) of the brandymaker, through the earthen cooling pots (*bucaros* and *alcarazas*) of the Portuguese, through the beds between which the poured wax candles cool, through scattering the malt mixture with chopped straw in the breweries, etc.

4 "Not so: these early inventions were thought out by no other class or men than those who have them in charge today" (Immo non aliis excogitate ista sunt quam quibus hodieque curantur). Seneca contradicts here an opinion expressed by Posidonius, that the wise man is the one who has discovered such things as the rudder of the ship and even the cobbler's trade. Seneca's point: it was not the "wise man," but rather the same class of man who uses them. Seneca, *Ad Lucilium Epistulae Morales*, trans. Richard M. Gummere, vol. 2 (William Heinemann; G. P. Putnam, 1920), 414–5.

It cannot be expected that in every handcraft the most advantageous means is already in use for every work or purpose. Instead, experience teaches us that from time to time improvements are done that make the work easier, more secure, or faster. How greatly would an aid to the so tedious, tiring, and unhealthy labor on the fulling cauldron be desired for hat makers.

Often a handcraft has been improved and perfected all at once, in that instead of the means that were exclusively used before for a work, another one is used, which another handcraft has long since employed for a similar purpose.

Before the sixteenth century all coins were stamped with the hammer: a dangerous, tedious, and slow process. At the same time, various artisans had long since given sheets of metal an embossed relief with the help of a roller or a press. What remarkable improvements the numismatic art received immediately when it occurred to someone to press the stamp into the metal sheets and plates with a roller, and better yet to use a large press with a long lever. Also, the striking mechanism (*Klipwerk*), whereby in recent times small coins are quickly and easily stamped, is from the iron stamping tool (*Wippe*) of the needle makers, with only slight adjustment.

That the processing of cotton could be greatly eased and improved if one, instead of using the cards of the wool weavers, would take the bow of the hat makers, proves the example of the Greeks and all oriental nations. That these nations first processed cotton, and have used and still use the bows they invented since the most ancient times, is certain: but I do not know what caused those who introduced the processing of cotton in Europe to take cards instead of bows. Perhaps they found these cards, whose use they had long known of, to be more comfortable. Presumably it is because they imagined to themselves that the removal of short wool (*fachen*) would have to be done, in oriental style, entirely while kneeling on the ground, whereas the wool-carder sits on his stool.

But how did it happen that one, in the case of hat-making, which is certainly a European invention, learned to apply right away the more advantageous – or rather, indispensable – tool, the bow of the oriental cotton-worker, and learned to do the removal of short wool comfortably while standing, by hooking the bow on a long thread, over a table punctured with holes? Only the Italians have recently begun to use the bow on cotton, and not to card, and the French have, according to those examples, also tried to introduce the bow in their cotton manufactures.

It would not be difficult to come up with more and much better examples that arts and handcrafts have been astonishingly improved through not only entirely new inventions, but also by the translation of means from one handcraft to the other. I borrow this expression from Bacon of Verulam,[5] who already ascertained that it could be more useful to the arts than anything else, if only the varied means already used by handcraftsmen for a single purpose could be made known to whoever had inclination, skill, and opportunity, to select the best of them and attempt their translation.[6]

One day, when the general technology will have been worked out, and when the inclination of the scholars for technological knowledge continues to grow, then many of them who will have time and opportunity will cause the translation of various means and tools to be attempted by talented masters.

It would naturally be best if those masters arranged the choice themselves. But this would occur most infrequently until more handcrafts, through study of auxiliary sciences and general technology, have been formed into general concepts.

Plato has said that the world would only be happy when kings became philosophers, or the philosophers became kings.[7] I don't know whether he also

---

5 Francis Bacon (1561–1626), Lord Verulam, philosopher and statesman.

6 (2) *De augmentis scientiar(um)*, vol. 2, p. 129. (; *quod*) *Nihil ad imrem quendam inventorum utilium, eorundemque novorum, veluti coelitus deducendum, tantum valere possit, quantum si experimenta complurium atrium mechanicarum uni homini, aut paucis, qui se invicem colloquiis acuere possint, in notitiam venerint; ut per hanc, quam dicimus, experimentorum translationem, artes se mutuo fovere, et veluti commixitone radiorum, accendere possint* (that nothing can so much conduce to the drawing down, as it were, from heaven, a whole showr (sic) of new and profitable Inventions, as this, that the experiments of many Mechanick Arts, may come to the knowledge of one man, or some few, who by mutual conference may whet and sharpen one another; that so by this, which we call *Translation of Experiments*, Arts may nourish, and as it were, by a commixture, and communication of Rayes, inflame one the other), Book 5, chapter 2, 148–9.

Lib. 2. cap. 2. at the end, where he speaks of *de usu historiae mechanicae* (of the use of the history of mechanics), which is almost the general technology, he says, what certainly applies to (the general technology): *Non* (*Neque enim ad praesens*) *tantum iuvabit, nectendo et transferendo observationes* (*unius*) *artis in usum aliarum, et inde novas commoditates eliciendo; quod necesse est fieri, cum experimenta diversarum artium in unius hominis observationem et considerationem venient; sed porro ad causas indagandas, et artium axiomata deducenda, lucidorem facem accendet, quam hactenus unquam affulsit* (For it will not only minister and suggest for the present many ingenious practices in all trades, by a connexion and transferring of the observations of one art to the use of another, when the experiences of several mysteries shall fall under the consideration of one man's mind; but further, it will give a more true and real illumination; concerning causes and axioms than is hitherto attained.) Bacon, *Advancement and Proficience of Learning*, 178.

7 "'Unless,' I said, 'the philosophers rule as kings or those now called kings and chiefs genuinely and adequately philosophize, and political power and philosophy coincide in the same place, while political power and philosophy coincide in the same place, while the many natures now making their way to either apart from the other are by necessity excluded, there

would have said this in the case of the single true philosophy, and that means: the currently fashionable one. I believe to be able to claim something similar about handcrafts, that these will only then have reached the greatest completion when scholars know handcrafts, will apply their knowledge to them, and when the handcraftsmen (the sons of the wealthy) will study those sciences which can serve them best, and the most honestly. Now various handcraftsmen are often seen working closely, side by side, without one particularly worrying about the method of the other. At least, most do not seem to have the idea that one handcraft could learn advantageous practices from another, or use the other's tools in its workshop to its own advantage.

And if this were also to occur to someone, then most of the German handcraftsmen can dare not at all, or only in small steps, to conduct such an experiment which could fail, at least in the beginning, and lead to a loss of time and expenses.

It is otherwise on that fortunate island, England, where the handcrafts are more honored and thus also pursued by noble, knowledgeable, and wealthy families. Where does one find greater and more numerous attempts made for the improvement of the arts than there? Where are new inventions better paid and more quickly implemented than there? Where do the handcrafts flourish more than there? The German recognizes this with envy, but an incorrect estimation of trades belongs to his original sin, which seems capable of no improvement.

A thousand examples prove how astonishingly much depends on the selection of means for the works in the workshops. There the German still spins on the spindle or on the wheel he invented. The Englishman lets children spin on the machines, so finely, that we can already buy yarn and fabric more inexpensively from England than we can produce it ourselves.

Soon, this will be the case with most wares. For who will want to work in competition with the English steam engine? From where would we get its furnace, which we can scarcely afford for our own warmth! Only the however manyeth German fabricant and manufacturer will be able pay for the steam engine!

The general technology that I propose here seems to me, in its organization and usage, to have great similarity with the systematic catalogues of remedies, the so-called *materia medica*. The *materia medica* mentions all medicines known up to now, the *simplicia*, of single effects, in as many particular sections as there are effects which can be identified. Löseke's *Materia medica* of

is no rest from ills for the cities, my dear Glaucon, nor I think for human kind.'" Plato, *The Republic of Plato*, 5:473, Allan Bloom, trans. (BasicBooks, 1991), 153.

Gmelin,[8] my deceased friend, first teaches all the known means of inducing vomiting, then particularly the means of purging. Thus too in a few paragraphs the means of inducing sweat, of providing strength, etc. It is truly a great convenience for the practicing doctor, who must naturally bring to the selection of medicine his own understanding, erudition, and experience.

When at some point, according to a similar method, the *materia technologica* (whose name I however do not wish to recommend) will have been worked out, then such a *materia technologia* will likewise ease the selection for the practical technologist who wishes to improve the arts. Still, he will need to possess understanding and various kinds of knowledge if he does not want to harm himself and others and make himself ridiculous.

It goes without saying that the means for the processing of materials according to this must be different based upon their nature. If tin and tobacco are to be made smaller (*zerkleinert*), who will then wish to stamp or grind the former or granulate the latter? Iron sheets unplated by tin are filed smoothly with sand. I once saw an inexperienced peasant maid rub silver spoons with sand because she had never learned of any other material for this purpose. The kiln for gunpowder has to be set up in a completely different way than a malt kiln.

But there are still various rules to observe which are not as generally known, or which do not occur immediately to everyone. These deserve to be collected, and placed in the introduction before technology. Meanwhile, they have still not been revealed, and I at least would not like to make the attempt to list them here. The following therefore only serves as an explication of what I mean.

Some means, as advantageous as they are for the larger facilities, where a few wares are constantly fabricated in quantity, are still too expensive and too inconvenient for those who, as one says, work on a smaller scale. The friezing-mill deserves a place where many curled rateens are required, but where a piece of this kind is only occasionally sold, one contents oneself with good reason with the hand-curler. At one location, where a few quintal of tobacco were to be processed, I found a taller and larger drying house than I have seen in Amersfort.[9] Many a schemer has harmed himself by transferring too quickly from the large to the small. On the other hand, perhaps even more

8 Johann Ludwig Leberecht Löseke's *Abhandlung von den auserlesenen Arzneymitteln* [Treatise on selected medications] was, in its sixth edition, expanded upon and edited by Johann Friedrich Gmelin (Friedrich Nicolai, 1790).

9 Amersfoort, Netherlands.

have become unhappy in that they wanted to apply what they found possible on a small scale on a larger one as well.

Some means usually employed by a handcraft could be applied just as advantageously to another material if they were not inconducive to its further processing. Polishing with the polishing stone can be made incredibly easier for various materials if the stone is rubbed with some mutton fat, but when paper is smoothed in this way, one cannot permit oneself the use of the fat because otherwise the paper does not accept ink. Through sulphur steam the dirty color can be removed from some wares and the white color increased, but cloth which has been embroidered or printed with gold or silver does not tolerate sulphur treatment.

Thus various means exhibit more than one good effect at the same time and are for that reason all the more advantageous. The beating of glued books and various kinds of leather do not just make these wares smoother, but also thicker and more durable, and the worker is happy about this. But often such means cannot be applied where the one effect might be desired, but the other cannot be tolerated. I wish to give a few examples.

Malt is dried more quickly and in some regards better in the kiln than in air. But the kilned malt yields brown beer and is also, because of the effect of the kiln, advantageous where one wishes to have brown beer. One otherwise would have to color the beer through heavily roasted barley, as occurs here. Where wheat beer is desired, one is better off taking air-dried malt and should not use the kiln because one cannot easily prevent its one effect of coloring beer.

One often applies an oil polish to things which have a shiny surface, but its use for painting makes this shine a defect, which one still does not know how to remove.

Our pharmacists have invented various correctives (*corrigentia*), which is the means whereby a disadvantageous and unpleasant quality can be removed, or made unharmful or unnoticeable. Correctives must also be invented for the technology.

In some cases various means cannot be applied because they work either too slowly or too quickly. A few years ago in our neighborhood, one changed the oven of a brandy maker according to a suggestion for saving wood. Wood was spared, but soon thereafter one tore out the economical oven again, because the distillation went slower and the evil excise tax must be paid according to the time which the lid of the alembick is outside of the excise office.

Soap and sheep dung mill (*walken*) quickly and with strength, but if the wares need time to be cleaned and loosened into felt, then one does not need both, but rather bean meal or clay, which works more slowly.

There are means which afford a great savings in time and effort, but deliver the wares more poorly. The quick roller only delivers coins with a weak stamp, all of which are bent and not round, as the coins printed in Anspach prove.

Some wonderful machines cannot be used where the power or space for them is lacking. Not every paper mill can accommodate a water press, even if they consider such things advantageous, because there is not sufficient water to drive it. If this was not lacking so often, mill works would be constructed much more.

There are processes which make a ware more complete than it can be made in any other way, but they require too much skill, too much caution, and are thus much more unsafe than if they could be used by many. Pouring delivers mirrors of remarkable size, beauty, and refinement, but at Neuhauß in Lower Austria, where one has achieved great skill in pouring, a mirror for the prince of Lichtenstein 120 inches long and 60 inches wide only took after the sixth pour. It is thus preferable to blow mirrors no more than 30 to 50 inches long, which are more likely to succeed.[10]

Every section or article of the general technology would have to, in my opinion, be comprised of two parts, the general and the particular. The former would have to indicate the common and special purposes of the works and means elaborated in the other parts, explain the principles upon which they rest, and otherwise succinctly teach what could serve the comprehension and judgement of the individual means and their selection for translating to other objects than the ones which they are currently used.

This would facilitate, for artists and handcraftsmen, the basic and general concepts of the objects which they work on and the usual methods for that, and in general permit an overview which could lead inventive minds to new, useful improvements.

Furthermore, for the general technology to provide the greatest possible use it is capable of, books must, if not always, then often, be listed in which one can find workings and tools described and portrayed.

In recent times there has been talk of many machines which are supposed to facilitate certain jobs: for example such ones which cut, stamp, drill, and polish, etc. These too should be mentioned alongside the works for which they are designated.

If one observes these numerous projects in the countless periodical writings, one should consider our age particularly inventive and fortunate, but alas!

10 J. A. Schultes, *Ausflüge nach dem Schneeberge* (Excursions to the Schneeberg) vol. 1 (J. V. Degen Verlag, 1802), 54. The Schneeberg is a mountain in Austria.

Only very few machines are useful for the thing they are supposed to be, and it is true what Poppe has said in the preface to his *Encyclopedia of Machinery*.[11]

One has to wonder about the impudence whereby unknowledgeable persons recommend their supposed inventions, just as one does about the impudence of those who translate descriptions of foreign projects without having the slightest knowledge of the objects. What is more, the illustrations are often made so small, or even engraved in the wrong direction, or incompletely, so that one cannot make the slightest use of them.

Also, of those machines which could really achieve what their inventors promise, only few come into use because the works for which they are designated occur only seldom, and their purchase is too inconvenient, arduous, and expensive for such rare cases.

This also holds true for many machines with which some think to facilitate agricultural works. A paring-plow has been proposed with which turf can comfortably be trimmed back, but it costs 6 to 7 Thaler, and if at some time the villager is required to do this job, and he doesn't have one available, before he is able to procure it he is able to cut back the turf just as well with a turf hoe. Before he procured the *Abziehpflug*, draining plough,[12] from England, he dug the necessary holes with spades.

It nevertheless remains true that some projects deserved to be used and that a systematic catalogue of them would be desired. One finds more than a few listed in Krünitz's *Encyclopedia*.[13]

The works of the collective handcrafts and arts, and the tools and machines belonging to them, are much too numerous to allow anyone to believe that he could be able to know and classify them all. Whoever wants to make the first attempt has to hope that connoisseurs of these objects come to his aid with advice and suggestions, and that at some point they will improve and complete his design, which cannot turn out any way except extremely incompletely.

11 Johann Poppe, *Encyclopädie des gesammten Maschinenwesens oder vollständiger Unterricht in der praktischen Mechanik und Maschinenlehre* [Encylopedia of the collected maschinery or complete instruction in practical mechanics and the theory of machines] Voß und Compagnie, 1803.

12 Beckmann refers to both the German term and the English translation.

13 Johann Georg Krünitz, *Oeconomische Encyclopädie, oder allgemeines System der Land-Haus- und Staats-Wirthschaft* (Economical encyclopedia, or general system of agricultural, house, and state economy). Volume 1, from Aa to Am. Berlin: Joachim Pauli, 1773. There were 242 volumes of this encyclopedia published through 1858.

How many incomplete attempts were made before Gesner's Latin dictionaries, which finally approach completion, became possible?[14]

In order to contribute a small coin (*Scherflein*) to this far-reaching endeavor, I dare to present a few articles which I have worked out, as far as it has been possible for me to do so. At least they will explain what I mean with a general technology, however incomplete and faultily organized they may be.

I wish namely to thank those who desire to judge and improve those articles and my entire design with a decency which honors them. Whatever those who take pleasure in using their wisdom to bring about the diminishment of others will provide that is of use, I will also use once it becomes known to me. But as I see it, thanks and retaliation towards such people would be equally impermissible.

In addition I will allow myself to say with Ausonius: *Non me fallit fore aliquem, qui hunc iocum nustrum acutis naribus, et caperata fronte condemnet; negetque me omnia quae huc pertinent, recte attigisse. Quem ego verum dicere fatebor. – Sed qualiscunque est, cogitet secum, quam multa de his non reperisset, si ibse quaesisset* (I do not forget either that there will be someone who with keen scent and furrowed brow will damn this jest of mine, and say that I have not touched on all the aspects which the numbers three and nine present. I will admit that he speaks truth, but deny its fairness. For if he is a good sort, let him consider that what I have left out has not been forgotten by me, but passed over. Next, whatever he is like, let him reflect how many of these instances he would not have found if he himself had been searching).[15]

Zerkleinern (Making Smaller)

How, or with what, the parts of the body are bound together, or the cause of their connection, is for the investigators of nature to look into. Until now, they have only presented hypotheses about it, among which not a single one

14 Johann Matthias Gesner (1691–1761). Beckmann quotes Gesner at length in his preface to Jacobsson's technological dictionary.

15 Beckmann edits the Latin text somewhat. The English translation is of the full passage. Below, the parts Beckmann keeps are italicized:

Neque *me fallit fore aliquem, qui hunc iocum nostrum acutis naribus et caperrata fronte condemnet negetque me omnia*, quae ad ternarium et novenarium numeros *pertinent, attigisse. quem ego verum dicere fatebor*, iuste, negabo. quippe si bonus est, quae omisi, non oblita mihi, sed praeterita existimet. dehinc *qualiscumque est, cogitet secum, quam multa de his non repperisset, si ipse quaesisset*. See *Ausonius*, "The Riddle of the Number Three," trans. Hugh Evelyn White (Heinemann; G. P. Putnam's, 1919), 356.

explains everything, and thus not a single one seems true. Here, the experience that all bodies can be disjoined or broken into smaller parts is sufficient.

Whether this division goes into the infinite, or in other words, whether every particle can always be divided again, this question, which even the investigators of nature leave to the metaphysicians, is not our concern here. For us, it suffices to know that ever new means are being invented, to push the division or refinement of the particles further than has ever been possible before, and we still do not know how far it can be pushed.

Therefore, where the finest divisions are desired, it continues to be a reasonable task for the spirit of invention to divide the particles even more – to prepare an ever finer flour, an ever finer sniffing tobacco, or finer pigments.

The force that is necessary for dividing is not the same for all bodies. Those which demand a greater force are called hard, and those where a lesser force is required are called soft.

Such bodies, whose particles can be given all manner of positions without the connection being broken, are called viscous, ductile, or malleable bodies. To those belong soft wax, soft clay, and the metals which can be hammered or stretched, which can be pounded into a sheet and pulled into a wire.

Bodies which are neither tough nor pliable are called brittle bodies. From these, several parts can be separated when force is directly applied to them. To these belong vessels of glass and clay and other fragile things.

There are bodies comprised of many threads, leaves, or flakes lying one atop of the other which, when the force is applied in the direction of the threads or layers, are easier to separate, or are much more brittle, than when one tries to break them in the opposed direction. Several of these bodies permit being broken into smaller leaves or threads, but not being pulverized or made into dust. The true Russian glass[16] and some other types of mica (*Glimmerarten*) can be broken down into extremely fine leaves. Asbestos allows itself to be easily broken into smaller threads, but I do not know whether there is a way of making either into the finest powder.

There are bodies which give way under pressure or bending and thereby allow for their shape to be changed. Some of these retain this shape, even when the force operating on them has stopped. Others, however, resume their old shape again, as soon as that which pressed or bent them stops. The latter are called elastic bodies.

Finally there are bodies whose parts can be separated with the least amount of force and which therefore take on the shape of the vessel in which they are

16 Also called Muscovy glass. The technical name is *lapis specularis*.

found. They are called fluid bodies and seem to require no division in that they are already comprised of the smallest particles. Nevertheless, the so-called killing of quicksilver, when it is rubbed with oily or slimy – at times dry, at times fluid – substances, can seem to be a division, because at least the smallest particles are thereby removed or kept apart from each other.[17]

All those bodies that are ground to the finest powder seem similar to fluid bodies, but they allow themselves to be piled above the rim of the vessel containing them, which is not possible, or only scarcely, with fluid bodies.

When there is doubt as to whether one should call a body hard, elastic, brittle, or soft, then one should remember that actually all bodies are hard and none are without any elasticity. They only differ according to the degree of hardness, of brittleness, of elasticity. According to these various differences of the bodies, one has to choose the most advantageous means of dividing them.

Brittle bodies are usually easier to crush, the drier they are. Dry rubber and very dried-out resin are easy to grind, but when damp they are both tough and do not crack.

In glass-like minerals that are very solid, the work becomes easier if one first anneals them and then cools them off in water. Often this must be repeated a few times. Thus glass-like stones become porcelain and authentic English earthenware, thus blue glass cools out of cobalt, in order to be pulverized more easily. That which becomes tough through heat, like some kinds of rubber and resin, can be more easily reduced to small pieces with great coldness.

Organic bodies that need to be separated into little fibers are first softened in water, often to a degree of rottenness, partially in order to dissolve the gummy material that holds the fibers together and to separate the fibers through the penetration of the water particles, and partially in order to make them more pliable. One therefore lets hemp and flax ret,[18] and the lumps for paper become rotten, although the latter is not very necessary.

With fibrous or tough bodies, dividing them is made easier if one adds to them harder, more brittle and fragile bodies and then breaks both into pieces at the same time. Then the harder pieces act upon the softer ones like wedges, separate them from one another, and ensure that the fibers are not transferred

17 "Certainly quicksilver is prevented from returning to its whole form, as it would otherwise do, by human saliva, or axle-grease, or turpentine, and suchlike, which prevent its parts from uniting because of their lack of agreement with bodies of this kind. When surrounded by them, they withdraw; and thus their avoidance of the intervening substances is stronger than their desire to unite with the parts that are like them; this is what they call the mortification of quicksilver" (Bacon, *New Organon*, 199).

18 To "ret" is to soak hemp or flax in water so that it softens and the fibers are easier to separate.

from one to the other and that they don't become felted, or that the separated (*zerkleinert*) pieces don't clump together, or allow them to become smooth. This method is used in apothecaries for crushing roots which in any case are often to be mixed as powders with other materials. Thus leaf-brass (*Goldschaum*) is pulverized with candied sugar to a fine powder in order to be able to apply it with brushes to glass and porcelain as gilding. So that the sugar does not at the same time harm it, however, it is sieved away afterwards from the gold powder with water, in which one can then apply it to further use.

One must take care that the divided body not become impurified by rubbed-off particles from the tools: that flour does not become impurified by the sand of the mill stones, medicines by the copper of the mortar, or pigments and glazes by rubbed-off particles of the mortar, stamps, and grinding stones. One therefore uses flint tablets for the latter.

At times, that which is supposed to be crushed can be wrapped in strong paper, and if impurification cannot be completely prevented, at least the powder can sometimes be purified again. For example, the glass powder can, by pouring on acid, be purified of chalky or metallic particles.

Whichever means of dividing one chooses, the powder will still never attain an equal degree of fineness. There will always be coarser and finer pieces. It becomes necessary to separate these through various sieves or through cleansing in order to sort the powder according to its fineness, and in order to separate the pieces which are harder to divide and, if necessary, work on them further.

With costly and dangerous materials, one must avoid pulverizing either by moistening with water or by covering them up. For example, in order to make diamond powder in the diamond mortar, in the processing of gun powder, and in unhealthy materials in apothecaries, where the pulverizing is safeguarded through the addition of some almonds. Occasionally, workers have to protect themselves from harmful dust with masks. It is worth the trouble to determine the loss incurred by the pulverizing of every piece of merchandise to make it as small as possible. See Westrumb on *Arzeney-Taxen* [Classifications of medicine], volume 1, p. 76.[19]

In order to maintain the necessary degree of fineness, various means must often be used one after the other. Wheat first becomes malt when expanded through germination, and the malt is then shredded. Some wares are first stamped, and then finely ground or pulverized. For example: quantities of

19 Johann Friedrich Westrumb (1751–1819) was a German apothecary.

varnish, or the leaves for snuff. Rags first become paste in the washing cylinder, then pulp in the beater.

The practiced artisan and merchant determine the degree of fineness by sight, touch, and according to the feeling between the teeth and on the tongue. A measure of fineness has clearly not yet been invented. In chemistry, one named the finest powder *alcohol.* This name, since Theophrastus Paracelsus's time,[20] is given to all solid and liquid materials if they are extremely tender and fine.

Some mixtures of many divided (*zerkleinerten*) bodies are baked together again, or otherwise shaped into a solid body, which must afterwards be separated again into small parts for use. For example, chocolate, Indian ink, pastel colors. In these cases it is often not easy to achieve the most desirable degree of firmness. Also, some of the tablets and pills of the apothecaries belong to this category.

All means of dividing bodies into smaller pieces which are used by the artisans, and which one calls mechanical, deliver only similar pieces which preserve the nature and quality of the divided body. The finest file-dust of the iron worker is iron. Every salt remains such, unchanged, no matter how finely it has been crushed.

But most bodies are composed of various parts which, most precisely joined, comprise the body, and can only be divided by chemical means. This separation (*Scheidung*)[21] or partitioning of the bodies into their basic materials or components is called disjoining (*Zerlegung*), decomposition (*Zersetzen*), or analysis (*decompositio*), whereby one must look for instruction in chemistry, which in several regards naturally comes so close to technology, so that the limits of each can scarcely be determined. Mechanical partitioning (*Zertheilung*) breaks bodies into smaller pieces, chemical partitioning destroys them.

The purpose for which bodies are broken into smaller pieces is usually, above all, in order to make them better suited for all effects that are stronger or weaker according to which the body has more or fewer surfaces, and is larger or smaller. The wind cannot blow away a cubic shoe[22] of marble, but it can turn it very easily into dust if (the marble) is broken into thousands upon thousands of small cubes and its surface is therefore multiplied a hundred times over.

20 Paracelsus (Theophrastus von Hohenheim) was a sixteenth-century Swiss physician and alchemist.

21 One can recall here the earlier name for chemistry, *Scheidekunst*, which is the art of dividing.

22 *Kubikschuh*, equivalent to a cubic foot.

In this regard bodies are divided (*zerkleinert*) which are to be dissolved in water, spirit of alcohol, or salt solutions, and those from which one wants to extract particular components. Thus one roughly grinds malt for beer, brandy, and vinegar. Thus one rasps down dye-woods, grinds coffee, and stamps down the hides for the tanner.

Also, many medicines become all the more strong, the more points there are touching the part of the body upon which they should have an effect. Thus snuff for noses which have become thick-skinned.

Dividing is also necessary in order to be able to mix several bodies uniformly, or in a particular proportion. For example, to be able to process sulphur, saltpeter and coal for gunpowder, flour for bread, spices and victuals for dishes, or all sorts of minerals for dishware.

In order to cover or coat one body evenly with another, the latter must be finely split, such as in the case of goldplating, decorative painting (*Stafirmalerey*), baking (*Conditorey*), velouté wallpapers (*bestäubten Tapeten*).

In order to be able to rub, scour, polish and smooth one body with another, the one that does the rubbing must sometimes be pulverized more coarsely, and sometimes more finely. Thus one powders, grinds, and washes (*schlämmet*)[23] emery, tripoli, pumice, and sandstone.

In order to undertake a separation, when various types of bodies of diverse specific weights, or with diverse shapes of their smallest pieces, are mixed, one makes them smaller. One therefore does this with ores, in order to cleanse and wash them, just as with grain, in order to separate the flour from the chaff (*Klei*) and to separate starch from the other components.

In order to prevent sticky things from baking together, one strews an appropriate powder on them. Thus pills are covered with witch-meal (*pulvis lycopodii*), or with pulverized licorice root, or with Dutch gold.[24] Thus in the cases where porcelain is fired, one strews sand upon the points of contact, in order to prevent them from melting together.

For the same purpose, to prevent attaching, things can occasionally be blackened or coated with the soot of burnt resinous wood or an oil lamp. For example, when pouring metals in an iron forge.

The means of separating bodies into only a few pieces, or into pieces of a particular quantity, size, and shape, are more appropriately collected under the concept and heading: carving off (*Abschneiden*), cutting up (*Zerschneiden*), or the latter perhaps under forming (*formen*) as well.

23 See Adelung (final edition), s.v. "schlämmen," where he defines it in terms of separating fine from course powder from a powdered body.

24 Dutch gold is an alloy of copper and zinc, used as an inexpensive imitation of gold leaf.

To the first case belongs the cutting up of glass with a diamond or flint, also with a pastil for fumigating (*Räucherkerze*); the springing (*Absprengen*) of alembics, phials, and other glasses with smoldering spring irons (*Sprengeisen*); and the cutting of soft bodies, like soap, cheese, butter, or clay with a wire.[25] To the latter case belong the many kinds of knife, scissors, pincers, the cutting through of coins, the stamps (*Stanzen*) of the flower-makers, girdle makers, the hollow punch (*emporte-pièce*)[26] etc.

In this category is also to be included the machine of an Englishman that splits skins and leather according to length and width. See *Repertory of arts and manufact*(*ures*) vol. 4, p. 104.[27]

I ***Brittle Bodies*** **(Spröde Körper)**

1 Striking to Pieces (*Zerschlagen*)

Thus ore pieces (*Stuferz*) are shattered with the hammer or iron in order to separate ore from ore-free (*taube*) stone. Often a wedge and anvil are used for striking. This happens with the splitting of the diamonds which the artisan Andreas Bevelmann in Amsterdam perfected and with the hitting of flint stones.

2 Chopping (*Zerhauen*)

For this purpose the various kinds of axes and hatchets. Some handcraftsmen use for this the word *Schroten*.[28]

3 Shattering, Crumbling (*Zerbrechen, Zerbröckeln*)

4 Crushing into Pieces (*Zerdrücken*)

For this purpose the use of sugar scissors,[29] also some screws and presses. For example, the common nutcracker.

25 Beckmann adds the synonym *Thondrahte*.

26 Technical term for a kind of paper-punching device.

27 There one can read about the "Specifications of the Patent granted to Mr. George Choumert, of Five-Foot Lane, Mermondsey-Street, in the County of Surrey, Tanner; for his Invention of a Machine for cutting, splitting, and dividing, Hides and Skins, both in the Pelt before dressed and when dressed into Leather, severing the Grain-side from the Flesh-side." Anonymous (T. Heptinstall, 1796), 104–106.

28 The verb *schroten* means to crack or crush into rough pieces.

29 "Sugar scissors, as the name suggests, were fashioned in the form of an X, with open, circular finger holes by which to grip and close the ends of the scissors around a lump of sugar. The scissor form became popular in the second quarter of the eighteenth century and remained so into the twentieth century." Sarah D. Coffin, "servers, sugar," in *The*

5 Stamping (*Zertreten*)

The stamping out of seeds by man and cattle, instead of threshing.

6 Breaking Down by Milling (*Zerwalzen*)

Thus starch is broken down to powder with a hand rolling mill. Also indigo and other pigments, as well as starch for powder and various materials of apothecaries are broken down with heavy stone or iron balls in mortars or kettles.

To this also belongs the mill-work, where a stone is attached to one arm of the axis resting on the lower mill-stone (*Bodenstein*), and is moved around by it (the axis) on the sharp edge, in order, for example, to break down the parts of gunpowder. So that the mass is not allowed to be smoothed flat, a shoe, or snare (*Schleife*), or rake is brought behind the runner, whereby the mass is stirred.

With a similar roller (*Walzwerke*), bark is now pulverized (*zermalmet*) for the tanners in our neighborhood, in Münden.

In Cheshire, England, bark is made smaller by rollers of cast iron. To these also belong the rollers of plowmen.

7 To Crush by Pounding (*Zerstoßen*)

With the pestle in mortars. Both have to be comprised of hard, fast, but not brittle materials, from bell metal, iron, rock-stone (*Felsstein*). Wooden ones are only seldom useful.

To prevent scattering (*zerstäuben*), the mortar is covered with a cloth or skin, or it has a cover with an opening in the middle for the pestle.

To ease the work, the pestle can be hung upon a mobile rod (*Schwingruthe, Streberuthe*),[30] which pulls it up after every strike.

Bodies which heat up through impact must be crushed slowly.

Bodies which ball together have to occasionally be stirred or rubbed rather than pounded.

8 Grinding (*Zerreiben*)

With a pestle in mortars of serpentine stone, agate, glass or with a grindle-stone (*Läufer*) on a grindstone of porphyry.

Oxford Companion to Sugar and Sweets, ed. Darra Goldstein (Oxford University Press, 2015), 601–603, 603.

30 Johann Heinrich Moritz von Poppe. *Ausführliche Volks-Gewerbs-Lehre oder allgemeine und besdondere Technologie* [*Complete theory of civil industry or general and specific technology*], (Hoffmann, 1839), 152.

9 Chipping, Filing (*Zerraspeln, Zerfeilen*)

As with rasps, so too with graters and files. Thus the dye-woods are filed down, mostly by prisoners in houses of correction.

Here, too, belong the various rasps of the tobacco manufactures, whereby the rolls (*Carotten*) are rubbed (*zerrieben*), rasped (*rapiert*) or made into rappy.[31] A few appear as a table-frame with saw blades, upon which the worker pulls one or more rolls back and forth at the same time. With other such tables the saw blades, as opposed to the rolls, are moved. The blades are then placed on a cylinder within a box, or in a horizontal frame, and in the former case turned with a crank, in the latter case pulled back and forth. Small graters belong to this category as well, whereby some of the rolls are smoothed down just before use.

The filings of various metals and metallic mixtures give – after they are plated, and after they have been set on a hot board – the so-called Nuremberg Powder-Brass (*Streusand*), which serves for all kinds of decorations.

10 Grinding (*Zermahlen*)

To this group belong the various grinding mills, which are moved by hand, cattle, water, wind, or by steam engines. The hand mills of the chemists, of which a few are described and depicted in Wallerius, *Physical Chemistry* I, p. 62.[32] The pepper mill, which was made to a coffee mill in the eighteenth century, to which the so-called 'Hollander' of the paper mills is similar. Also, the corn mills, spice mills, the mills of the blue dye fabric, and the enamel mill (*Glasurmühle*) of the porcelain manufactures and earthenware manufactures (*Fajancerien*). The construction of such things is taught by the art of mill building, for example Beyer's *theatrum machinrum molarinarum*. Dresden 1735. Fol. And again in 1767, to which Weinhold in Dresden 1788 (fol.) delivered another volume. See also (Johann Beckmann's) *physikalisch-ökonomische Bibliothek* (*physical-Economical Library*) XV, p. 569. Above all there belongs here H(einz) G(ustav) Flörke *On Mill Construction and Milling* (*Vom Mühlenbau und Mühlenwesen*). Berlin two parts in octavo. = Krünitz *Encyclopedia*, parts 95 and 96.

11 Sawing (*Zersägen*)[33]

The debris from saw mills, the sawdust, is used for packing, for drying off needles, for planchets (*Münzplatten*), and also for rough packing paper. In order

31 Rappy is a kind of course snuff.

32 Johann Wallerius, *Die physische Chemie* [The physical chemistry]. Part one. (Christian Mevius, 1761).

33 As opposed to *sägen*, *zersägen* connotes sawing 'down' to create something (planks, etc.).

to gather them neatly, I have seen in Holland a bellows placed under a wagon, which blew the sawdust into a crate after every cut.

The art of mill construction teaches how to build saw mills. To the most recent improvements belong the finer saw blades, which deliver less sawdust but smoother boards. See *Forstmagazin* (Forestry magazine) volume 5, p. 102. Also the clamping of multiple blades together. (Johann Georg) Büsch observed a mill in Sweden in which a water wheel drove 72 saw blades. See his *Reise durch Schweden* (*Journey through Sweden*), p. 14.

To the improvements of hand saws belongs: J(ohann) C(hristian) Schäffer's *Abbildung und Beschreibung einer dreyfachen Sägemaschine* (Drawing and description of a three-part sawing machine). Regensburg 1769, quarto.

12 Planing (*Zerhobeln*)

Where bodies are cut down through pushing. To this group belong the various planes of the joiners, carpenters, the basketmaker and sievemaker, the organ builder, the book binder, and others. See Krünitz, *Encyclopädie* vol. 24, p. 22. To these one can add the plane with which cabbage and some turnips are sliced.

The fine shavings of the joiners serve for packing, also as rough packing paper. Rather than with the usual rasp, dye-wood is planed with a long one which has various planing irons in the casing. More fine shavings are thereby made with less difficulty, in shorter time. One can consult the illustration in (Zacharias Conrad) von Uffenbach, *Reisen* (Travels), vol. 2, p. 92.[34]

13 Separate by Turning (*Zerdrehen*)

Thus pure tin is turned into fine strips for dyers, for the so-called composition, in the case of scarlet, king's yellow, and other colors.

14 Biting, Chewing (*Zerbeissen, Zerkauen*)

One will kindly allow me, even if a smile can't be helped, to include this oldest, and for animals most natural milling and grinding device, to which especially belongs the disgusting preparation of the drinks of the wild Americans, which is called *Chica*, from seeds and herbs chewed by old women and prepared with their spit for fermentation. One may consult Goguer, *Ursprung der Gesetze und Künste* (Origin of the laws and crafts), vol. 1, p. 111.[35] *Algemeine* (sic) *Geschichte*

---

34 Zacharias Conrad von Uffenbach, *Merkwürdige Reisen durch Niedersachsen Holland und Engelland* [Noteworthy travels through lower Saxony, Holland, and England]. (No publisher listed) Frankfurt und Leipzig, 1753.

35 Antoine-Yves Goguet (1716–1758) was a French jurisconsult. The work was originally published in French as *Origine des lois, des arts, et des sciences, chez les anciens peuples* (Origin of laws, arts, and sciences among ancient people). In three volumes, quarto, Paris, 1758.

von Amerika (General history of America), vol. 1, p. 331. (*Herman*) *Boerhaave*[36] *praelectiones* (lectures), Haller's edition, vol. 1, p. 179.

15 Splitting, Tearing (*Zerspalten, Reissen*)

Wood is comprised of fibers or filaments which can be more easily separated lengthwise than in the opposite direction. The former is called splitting, the latter breaking.

Splitting has advantages over sawing. The work goes faster. Split or torn things are more durable in that the split follows the direction of the fibers without tearing any of them, as sawing does. They can more easily be bent for hoops, boxes, and other things. Also, several pieces can be torn, as opposed to sawed, from a clump, because the saw takes away a stroke of two, often three lines,[37] since the staves are at most only three lines thick. Also, clumps can be cut from some tree trunks that are not suitable for construction, from which all sorts of things can be pulled off.

Freshly cut wood often splits better than dried out wood, which has become much too brittle.

Among pulled wares belong: wine posts, oars, pickets, pipe staves, saddle wood, rims for sieves, box wood, book shavings (*Bücherspähne*), shavings (*Spähne*) for dagger blades, etc.

One finds splitting with the hatchet, along with the other tools necessary for it, described and depicted in Duhamel, *Von Fällung der Wälder* (On the cutting down of forests). Nuremberg 1766. Quarto.[38]

To the artistic works belong the fine sieve rims and box boards (*Schachtelbretter*), which are evened out on a particular frame with a cutting knife.

Even more artfully, the finest and longest shavings, often only one quarter line (*Linie*) thick, 3 ½ shoe (*Schuh*, meaning foot) long and 6, 12, up to 14 inches (*Zoll*) wide, are pulled with a particular plane which is pushed with a handle by men, or by a waterwheel. One should see Duhamel, table 29, figures 1, 2, and C. J. Löscher, *Kurzer Unterricht von Spähnkolben* (Brief instruction in wooden pistons). Leipzig 1783. Octavo. See (Beckmann's) *physikalisch-ökonomische Bibliothek* (*Physical-economical library*) vol. 13, p. 125.

Also belonging here are the whip handles split and braided – up to the grip, or in twenty rods – from alder or field maple, which are made by a particular society of tradesmen in a few Eisenach villages, also in Ohrdorf in the district of Gotha, and sold in great quantity at the Leipzig trade fair. A complete

36 Herman Boerhaave (1658–1738) was a Dutch botanist and chemist.

37 Eger's *Technological Dictionary* defines a *Linie* as 1/10 or 1/12 of an inch (*Zoll*).

38 Henri-Louis Duhamel du Monceau (1700–1782) was a French physician and engineer.

description of this art is still not known to me. In the meantime one finds the guild articles for the Kaltennortheim Office from 1733 in Schreber's *Neuen Cameralschriften* (New writings on cameralism) 16, p. 164, and on p. 112 of the same volume the guild article for the Lichtenberg office for the year 1737.

In France similar handles are made from the hackberry tree, *celtis australis*, which is called *micocoulier*, or *bois de Perpignan*, and in Perpignan *fabre coulier*. It frequently grows in Roussillon, above all around Ceret.

In a similar fashion the so-called Bengal cane (*spanisches Rohr*) is split with a knife by chair makers, first in four, and then every quarter in two, and for fine work into four additional parts, so sixteen altogether.

Here too belongs the pulling of fish bones, since the beards of whales, after they have been boiled, cleaned, and scrubbed are torn into strips of various strength. Through drying and aging it becomes brittle and breaks on its own.

II *Soft Bodies* (Weiche Körper)

16 Cutting (*Zerschneiden*)

The breaking up of rags with the so-called rag-cutting machine (*Lumpenschneider*). The cutting of straw on the chopping bench or chaff cutter, or on the newly invented straw mills. The chaff cutter is like the cutting machine where tobacco is cut, which is also driven by water. The waste from fabric scissors, sheering wool, sheering hair, serves for upholstery padding and velouté wallpaper.

17 Stamping (*Zerstampfen*)

Tobacco, vegetables and other things in containers whose iron is at times shaped like a wedge, at time shaped like an *S*. To these belong the pounding mills: bark mill (for tanners), madder mill, paper mill.

18 Hacking (*Zerhacken*)

With the hacking knife of the butcher, of cooks, etc.

III *Fibrous Bodies* (Faserichte Körper)

19 The Breaking of Flax and Hemp (*Das Brechen des Flachses und Hanfs*)

On the kneading bench (*Brechbank*), which is depicted in Zink's *ökonomischem Lexicon*[39] (Economic lexicon), table 2, figure 1, and in Krünitz's *Encyclopedia*.

39 Georg Heinrich Zincke (1692–1769) was a German jurist and cameralist. The reference is to his *Allgemeines oeconomisches Lexicon* [General economic lexicon] (Gleditsch, 1764).

20 Hackling[40] (*Das Hecheln*)

Where flax and hemp are pulled by the pointy, rough, or fine spikes of hackles which are fastened upon benches: in part, to separate the fibers from each other, in part, in order to cut the longer ones from the shorter ones, the oakum.

21 Carding (*Das Kardetschen*)

The clothmaker, stocking-knitter, hatmaker, wigmaker, wadding-maker, etc., with the various kinds of cards, wool-combs, curry-combs, scrapers, fine cards, and with the combs with the long double teeth, whereby one's purpose is the loosening and cutting away of short fibers, the scraps from the combed wool.

22 Loosening (*Das Flacken*)[41]

Either beating the wool with small sticks on wooden or iron hurdle, or on a net. *Flacka, flaka,* means splitting in Swedish and Danish. *Flacka fiskar* means to split or tear apart fish, whereby the term *Flackhering* or, uncorrectly, *Flickhering* is derived.

23 Beating (*Das Fachen*)

With the hatter's bow, which was first invented for cotton, for which purpose it is still used in the Levant.

24 The Woof (*Der Wolf*)

A box, wherein the wool is pulled apart and twined together with a toothed winder and by the hooks attached to the inner walls of the box. The machine also serves in the cloth manufactures to clean dust from wool, and to mix variously dyed wool evenly. One finds a description and illustration in Jacobson (sic) *Schauplatz der Zeugmanufakturen* [Theater of cloth manufactures], vol. 2, p. 64, table 1, figure 1. One knows of some variations of this machine in England, under the name gigging-mills, towing-mills, machines for twitching wool. One of the best machines of this kind is described and illustrated in *Transactions of the Society for the Encouragement of Arts*, vol. 7, p. 193.

---

40 This is the operation of combing linen.

41 According to Adelung's dictionary, *flacken* and *flocken* are synonyms. In Ebers' dictionary the verb *flocken* means "to beat into flocks," where flocks, in this context are described as little tufts of wool or silk. Ebers, vol. 1 (1796), s.v. "flocken."

25 Tearing Apart (*Zerzupfen*)

Preparation of the *charpie*[42] from old linen, for surgeons' bandages. From dyed finely torn threads pulled from silken fabric artful pictures are made, according to the technique of stuffing or with tufted paper (*Streuarbeit*), and kept under glass.

IV *Metals*

26 Corning, Granulating (*Körnen, Granuliren*)

One understands thereunder the means of transforming metallic substances into grains or granules.

Because metals, due to their malleability, cannot be crushed by pounding, and because filing down is wearisome and difficult, and also contaminates the filings with little pieces of iron, one melts them, divides them into small drops, and allows them to harden by becoming cold.

Corning occurs either in water or outside of water: wet and dry corning.

Lead and tin can be corned *dryly*, in a dry, warmed tray wiped over well with chalk or reddle.[43] Once the metal has been poured in, one wraps the canister in a cloth in order to protect the hand from the metal that oozes through the cover, and shakes it long enough so that, instead of the shaking movement of the fluid metal, a rattling sound emerges.

Melted lead can also be corned in an iron mortar, in which one stirs it long enough with an iron pestle for it to stand firm. One is thus spared wiping or washing away and the drying of the corns in order to clean them of chalk and reddle.

Wet corning is necessary with bismuth, brass, and zinc, metals that become so hot with melting that wooden containers would burn. One pours them into a container full of water which is stirred with a rod, or into a container of lukewarm water, in which a cylinder lined with broom slips (*Besenreisern*), a granulating cylinder, or a corning cylinder, is turned.

One separates the small grains from the larger ones with sieves. In order to avoid danger, one should use the rules of caution that Cramer has listed in *Anfangsgründen der Metallurgie* [First principles of metallurgy], vol. 1, page 59 and page 146, where one also finds the necessary equipment illustrated.

---

42 Beckmann uses the French spelling, but *Carpie* was also used in German. A *charpie* is a bundle of threads torn from a piece of linen and placed on top of a wound before bandaging.

43 Beckmann uses *Kreite* instead of *Kreide* for chalk, which Zedler accepts as a variation. Reddle, red earth, is a translation of *röthel*.

The largest granulating fabrick is certainly the one in the Harz, where granulated iron has been used since 1772 in the high ovens for the melting of silver and lead ore. One permits several quintals of iron that have been melted in the high oven to be poured, in a drain on the ground made of coals, into a great vessel that has been sunk into the earth, plated with iron, and filled with water, in which one constantly stirs with an iron hook in order to split the iron into smaller pieces. The iron is thereby divided into countless grains ranging from the size of fine hail to half-inch pieces. See Gatterer's *Beschreibung des Harzes* (Description of the Harz), vol. 1, or *Anleitung den Harz zu bereisen* [Guide for traveling in the Harz], vol. 4, page 86.[44]

Here belongs as well the fabrication of lead shot (*Bleyschrot*) or shot (*Schießhagel*), as is typical at Osterode. One lets the lead melt in the most gentle fire, shakes pulverized orpiment (*Auripigment*, yellow arsenic) upon it, and pours everything into a pot filled with holes, like a sieve, which rests above a vessel willed with water. The grains (of lead) are strained, and also smoothed with in a scrubbing barrel with roughly crushed plumbago (*Wasserbley*).[45] See *Encyclopedia*, planches vol. 4. Tab. Fonte de la dragée et du plomb à giboyer. Krünitz Encyclopädie. Vol. 5. Page 701.

In Paris one also makes, I am not sure how, a silver-colored shot which does not dirty the hands. It is called *plomb italien* or *plomb blanc*.

But much better is the *patent shot* made in London, in the year 1760, at the manufacture established by Watts. The melted metal falls down in a tower 123 feet 6 inches high, before it reaches the water, and thereby attains the complete spherical form.

Very different from that preparation is the one which should produce the so-called ground shot (*gemahlenen Hagel*). In this process thin leaden tablets should be made into narrow strips, cut into small dice, and made into spheres within a great drum, which is turned in a frame with a handle. See Timäus on munitions fabricks. Leipzig. 1792. 8.[46]

In order to granulate copper for brass forges, one allows it to fall in fine drops into a vessel through an iron colander, wherein the water is moved by the turning of a winged cylinder. See *Traité de la fonte des mines par le feu du charbon de terre* [Treatise on the source of mines by fire of the earth's

44 Christian Wilhelm Jacob Gatterer (1759–1838), German cameralist and nature historian.

45 Graphite.

46 Gebhard Erich Leopold Timaeus, *Versuch über Gewehrfabriken, die Schießkunst und das Jagdwesen* (Reinicke, 1792).

carbon] by *de Genssane*.[47] Paris 1770. 1776. 2 volumes in quarto. Volume II, page 93. Jägerschmid, *Bemerkungen über die Fabriken der Graffschaft Mark* [Observations on a few metal fabricks in the Mark earldom] 1788. 8. Page 49. Table 3.[48]

It is often sufficient to separate metals, for example zinc, just into small thin sheets. To this end, one lets small drops of melted metals fall upon a cold stone plate smeared with tallow and cool down.

The making of wax into ribbons for bleaching has the greatest similarity with granulating. This too is melted and then falls in small drops through a colander upon a wooden shaft which is turned in such a manner over a vessel filled with water, that half of it is constantly submerged in water, and allows the drops that have fallen upon it to be formed into shavings and fall down. One can refer here to the illustration in the *Pariser Schauplatz der Künste* [Parisian theater of the arts], part 2.[49]

Before the invention of this granulating machine, which should also be useful for lead and tin, one formed wax into ribbons with a ball alternatively dipped in water and melted wax, or used a plate instead. See Hofmann's *ökonomische Chemie* [Economical chemistry], page 82.

27 Precipitation (*Präcipitiren*)

Metals can be dissolved into acids or in quicksilver (*Quecksilber*), and thereupon through precipitation (*Niederschlagen*) separated again from acids, and by evaporation (*Abdampfen*) from quicksilver, since they are thereby separated into the finest particles, which also have a completely equal degree of fineness. Chemistry gives the necessary instruction for this.

To this belongs the dissolution of gold in aqua regia (*Goldscheidewasser*) made with salmiac (*Salmiak*),[50] and its precipitation through fire-resistant alkali or through green vitriol.[51] That fine gold powder obtained thereby serves for gilding porcelain and painting in enamel.

---

47 Antoine de Genssane (*c.*1738–*c.*1780) was a French naturalist.

48 E. A. Jägerschmid. *Bemerkungen über einige metallische Fabriken der Graffschaft Mark* [Observations concerning a few metal fabricks in the earldom Mark] (J. G. Müller the elder, 1788).

49 Daniel Schreber. *Schauplatz der Künste und Handwerke, oder vollständige Beschreibung derselben, verfertiget oder gebilliget von denen Herren der Academie der Wissenschaften zu Paris* [Theater of arts and handcrafts or their complete description completed or approved by the members of the Paris academy of sciences] (1762–1805).

50 Ammonium chloride.

51 Crystalline ferrous sulfate.

Also, the so-called calcined gold (*Goldkalk*), or the precipitate of Cassius, gold-purple,[52] mineral purple, which serves for ruby glass, is the finely-divided gold precipitated from *aqua regis* through the tin solution.

For cold gilding, small lumps are dunked into a tincture of gold, dried, and then immediately burned to ash, in which gold is divided most finely. One rubs this ash on the silver with a cloth or cork moistened in water or saltwater, which should become gilded, to which the gold particles cling. One washes the silver, which sees almost no gilding, but appears by polishing with a burnishing stone.

For fire gilding, gold is dissolved in quicksilver and with this gold amalgam the metal, which should be gilded, is coated, whereupon the quicksilver brought to evaporation by fire, and the gilded piece is polished with a burnishing stone or with the polishing iron. See Lewis, *Zusammenhang der Künste* [Organization of the arts], part 1.

Silvering occurs in almost exactly the same way. For cold silvering one takes the silver that, either with copper or salts, has been precipitated from the salpetre acid (*Salpetersäure*),[53] desalinated (*ausgesüßte*) with water, and finely rubbed together (*zusammengerieben*) with tartar (*Weinstein*), cooking salt or salmiac, or even with alum. See Wallerius, *physische Chemie* [Physical chemistry], volume 3, pages 451 and 456.

The copperplating (*Verkupferung*) of iron dipped into a copper solution also belongs here, which the count of Saluces has proposed in *Melanges* (sic) *de philosophie et de mathematique de la societé de Turin* [Selections of philosophy and mathematics from the society of Turin]. Years 1770–1773, page 191.

Perhaps it is not superfluous in this context to recall here the fine, white, light magnesia which our apothecaries obtain from England. It is slowly precipitated with dissolved tartar (*Weinsteinsalz*) from a very diluted solution of acrid salt (*Bittersalzes*).[54] For all precipitates turn out more finely, the more the solution is diluted.

V *Salze (Salze)*

28 Dissolution (*Auflösung*)

All salts dissolve in water, whereby their parts are spread everywhere equally in it. The dissolution is therefore a division (*Zertheilung*) which, if one pours in more and more water, can be increased further than with all mechanical means, until an indeterminable degree of fineness (*Feinheit*).

---

52 A pigment formed by the reaction of gold salts with tin chloride.

53 Aqua fortis, or nitric acid.

54 Epsom salt.

In the same way, every rubber in water, every resin in alcohol (*Weingeist*), and every metal in a salt solution can be spread or divided to an indeterminable degree.

29 Efflorescence (*Verwitterung*)

If salt crystals are exposed long enough to open air, they fall into a fine powder or meal (*Mehl*), without having their essential qualities changed thereby. The loss of crystallizing is accompanied by the loss of water necessary for it, thus a half ounce (*Loth*) of decayed salt is more effective than a half ounce of its crystals.

This decay, or efflorescence, can be furthered with heat. Thus in Goslar one allows white vitriol (*weissen Vitriol*)[55] to decay to a fine meal in a kettle over the fire before selling it.

Further, one can also calculate the decrepitation (*Decrepitation*)[56] of some salts, primarily kitchen salt (*Küchensalz*), which bursts into small pieces with a noise when it is heated quickly. Without doubt this stems from the water and the air which is found in the crystals. One thus tends to let rock salt (*Steinsalz*), and also coarse brine salt (*Solensalz, Sohle-, Soole-*) crackle (*zerknistern*) before use.

In order to keep cooking salt fine-grained, the brine (*Sole*) has to be boiled quickly and strongly, and stirred a lot beneath the precipitate (*Soggen*). This also holds true for other salts, for example of bitter salts (*Bittersalze*), Glauber's salts (*Wundersalze*),[57] etc.

VI *Water* (Wasser)

30 Evaporation of Salt Water (*Gradiren*)

In order to strengthen the salt brines (*Salzsolen*), that is, to free them from a part of their excessive water, so that they rise (*anschiessen*) with less heat and in less time, one exposes them to evaporation, and because it occurs on the surface of the water, one thus separates the brine (*Sole*) in the graduation walls into an innumerable quantity of drops or spreads them on the gradation roofs over a great surface.

31 Spraying (*Versprützen*)

Division of fluids through spraying with a rod, a faggot of brushwood (*Quaft*), or with a sprinkler, a gardener's watering can.

55 Crystalline zinc sulfate.

56 Decrepitation refers to the audible disintegration of a substance through heating.

57 A decahydrate form of sodium sulfate.

Burnishing, Smoothing, Shining, Polishing (*Glätten, Schlichten, Glänzen, Poliren*)

Bodies which have no raised or lowered points on them are called smooth. There are no completely smooth bodies. The greatest smoothness is found on a few natural bodies, if they have not undergone any change. Among the smoothest are ice and several which have emerged through crystallization.

Liquid bodies, at least those of the finest kind, also appear smooth on their surface. Various surfaces which have to be smooth are therefore covered with oil or water. Bodies only differ in their degree of smoothness. Such bodies which have noticeable raised or lowered points are called rough.

More dense bodies and those which have a greater specific weight are capable of a greater smoothness, thus precious stones, crystals, agates, jasper, metals and dense, very heavy types of wood, such as guaiacum (*Pockholz*), *lignum guaiacum*, acquire an exemplary smoothness.

Soft, tender, brittle, and light bodies remain rougher, like sponge, chalk, pumice, and slate.

Flattening (*Schlichten*) is what one calls it when bodies which can be bent, folded and curled are made even in such a way that little or no traces remain of the bending, folding, and curling. Flattening (*Schlichten*) in the meantime has the more general meaning of smoothing (*Glätten*).

Bodies are smoothed in order to give some of them a more pleasant appearance.

Many bodies are smoothed in order to reduce rubbing, and thus to facilitate movement. The movement of swimming bodies is so easy because they are exposed to less rubbing.

Some bodies are smoothed to make them more permanent. Bodies which have many raised points are more easily gripped, caught, and torn. Through rubbing, parts are rubbed off, and bodies thereby become thinned and used up. Through smoothing, the parts – at least at the surface – are brought together so thickly and firmly that they do not allow themselves to be so easily separated afterwards.

Very smooth bodies are not wettened (*durchgenetzt*) very easily, because they offer the water fewer indentations (*Vertiefungen*) and opening.

Even very smooth bodies will be affected sooner or later by salt solutions and by vapors (*Dünsten*). Acids dissolve flat calcspar (*Kalkspath*), and teeth that still have their enamel (*Schmelz*), less quickly than common limestone (*Kalkstein*). One can compare strongly smoothed paper with blotting paper (*Löschpapier*), as well as the so-called watertight cloths and fabrics.

For precisely this reason, smooth bodies stay more pure and allow themselves, when they are dusty or otherwise dirtied, to be made clean again easily. Rough bodies, which have been very dusty for a long time, can almost not be entirely cleaned again. Rainwater runs from glazed roof tiles most purely.

Because smooth bodies rub away less, to reduce dust in rooms, floors, tables, stools, and cupboards polished with wax (*gebohnt*).

Smooth bodies are also protected more than others against being affected and dirtied by insects, worms, and various pests. Mice and insects, which do not have clinging feet like young flies and spiders, cannot find their balance on flat surfaces.

When things should not rub against each other, when delicate wares should rub their covering, then it must be smoothed. In this regard, Dutch gold (*Goldschaum*) is placed between paper wiped with fine iron clay (*Bolus*).

To bodies which one wishes to secure against drying out or evaporation, one gives a smooth crust, for example bread and cheese.

Bodies which either become mirrors or should retain a mirror shine have to be smoothed as much as possible. Thus only the firmest bodies are suited to be mirrors, and among them only those which have either the whitest color, or none at all.

Objects which should not be distinguishable by touch have to have the same smoothness everywhere. For example, playing cards.

In order to smooth bodies, the raised portions are often taken away by planing, shaving, filing, rubbing, and burning.

Or the raised portions are pushed down by pressing, hitting, and rolling.

Or the depressed parts are filled with an appropriate substance, for example with wax.

Bodies which are too loose or soft for smoothing are first soaked with a substance that can harden, in order to fil up the in-between spaces (*Zwischenräume*), as one does with the bleaching (*Stärken*) of wax, the smoothing of various fabrics, and the planing or liming (*leimen*) of paper.

Bodies which can be softened by dampening and warmth are moistened before smoothing and then smoothed warm. Thus the laundry will be somewhat moistened or sprayed again (*eingesprengt*) in order to be smoothed better with the hot iron. Thus fabrics (*Tücher*) are ironed hot, chintz (*Zitze*) and calico (*Kattune*) are rolled hot or calandered.[58] Thus the paper maker's press works best when the paper is still somewhat damp.

58 "Calander" is a technical term for pressing.

Bodies which become liquid through heat acquire an even greater smoothness, depending on how smooth the surface is where they harden by cooling. Thus the mirror tables (*Spiegeltafel*) are poured on a flat metal plate. Thus lights become smoother in glassy forms.

Bodies of this kind are smoother if one exposes their surface to a degree of heat which can only melt the uppermost part, for example sticks of sealing wax.

Bodies which due to their nature cannot acquire the necessary degree of smoothness are often coated or covered with another which is capable of more smoothness. They are occasionally varnished or laquered, or glazed, or plated, tinned, silvered, golded with metal. The Russian icebergs maintain the greatest shine in that the steep sides are sprayed with water and then allowed to freeze to ice once again.

Things which should preserve the finest polish, for example optical glasses, have to be touched or grasped as little as possible, and guarded as much as possible against dampness, vapor, and dust, and at most only cleaned by blowing or cleaned with a soft brush. Fine polished steel items should also during the winter not often be brought out of the cold into the warmth.

1 Planing Off (*Abhobeln*)
With the smoothing plane (*Schlichthobel, Gläthobel* (sic), also *Glatthobel*)

2 Shaving/Scraping Off (*Abschaben*)
Thus lattens (*Latune*, yellow brass) are taken away with the scraping knife on the scraping block (*Schabeblock*), as well as the wooden wares of carpenters with scrapers (*Ziehklinge*).

3 Filing Off
With smoothing files (*Glätfeilen, Schlichtfeilen*) of various fineness, which, if they are moistened with water or oil, refine themselves during the work, in that the rubbed off dust fills up the cuts (*Feilenhiebe*) more and more and weakens the efficacy of the file. Wood and other soft bodies are worked with fish skin and horsetail (*Equisetum*) for the same purpose.

4 Lathe Turning
As the lathe operators with the smoothing chisel.

5 Scouring, Rubbing Off (*Abscheuern, Abreiben*)
Diamonds with diamond powder, diamond dust or other precious stones, and metals, and glass with sifted sand (*geschlämmetem Sande*), with emery putty

(*Schmirgel*), tripoli (*Tripel*) which, if it is too soft, is first burned with chalk, pumice, ochre, calcothor (red fine grained hematite), hematite (*Blutstein*), tin ash (*stannic oxide*), potash (*Pote*). Coin planchets are smoothed on each other with charcoal powder (*Kohlengestübe*) in the scrubbing barrel (*Scheuertonne*), just like shot (*Schießhagel*) with black lead (*Reißblei*). On the silk reel (*Seidenhaspel*) threads become smoother by crossing over (*Kreuzen*), in that one thread rubs upon the other.

6 Grinding Off (*Abschleifen*)

On the grinding mill (*Schleifmühle*), which is operated by hand, foot, or water. It either has a sandstone (*Sandstein*) or a smooth or chisel-shaped iron or steel disk (*Scheibe*).

Some wares retain the final polish on a wooden disk that has been covered by a piece of leather smeared with ochre. Among these number the grinding mills of the glass grinders.

7 Sleeking (*Glatstreichen*)

Pottery (*Töpferwaaren*) and breads (*Brode*) are, once they have been somewhat dried by the wind, smoothed flat with a wet hand.

8 Burning Away (*Absengen*)

Stockings which should be smooth are, if stretched on forms in a burning fire, robbed of their protruding threads, just like the bee hives (i.e., beehive shaped baskets) woven of straw.

9 Striking (*Schlagen*)

The striking of the paper formerly on the so-called stamping mill. The striking of the book binders. The flattener (*Schlichthammer*), sleeking hammer (*Glanzhammer*) of the tinsmiths (*Blechschmiede* and *Klempner*).

10 Pressing (*Pressen*)

Cold and hot pressing of still somewhat damp cloths. The pressing of paper, of books.

11 Rolling (*Walzen*)

Some fabrics and ribbons are smoothed by wooden or steel rollers (*Walzen*). One often places a hot iron (*Bolzen*) in the latter. To these belong the fabric rolls and large cylinders, the rolling of tin, of paper. The rollers of dough in kitchens and bakeries, also of clay. The rolling of garden paths.

12 Polishing by Hand (*Poliren mit der Hand*)

With the smoothing stone (*Glätstein*), often of glass, more often from agate (*Achat*), flint stone (*Feuerstein*), jasper, which is sometimes fastened to a polishing rod (*Glätstange*) and works better, the smaller the contact surface is, the stronger the pressure is, and the smoother the base (*Unterlage*) is, often in the shape of a half cylinder. Thus calicos (*Cattune*), fine linen, formerly paper, polished pasteboard, pressboard of the cloth dressers (*Tuchbereiter*), and playing cards are smoothed. Almost in the same way saffians[59] and other leathers are processed with the smoothing club by cobblers with the smoothing wood, and with the smoothing board (*Glätschiene*). Book binders and gilders (*Vergolder*) need the polishing tooth (*Glättzahn*), the polishing club, and the polishing steel.

13 Polishing with Smoothing Machines (*Poliren mit Glätmaschinen*)

Here, the best ones have to be noted (*verzeichnet*), with advertising (*Anzeigung*) of their descriptions and illustrations. One finds some mentioned in Poppe's *Encyclopädie des Maschinenwesens* (Encyclopedia of machinery), vol. 2, page 597. *Journal des mines* (Mining journal). Year IV, number 18, and Baillet's *Anmerkungen über die Marmorglätmaschinen* (Observations on machines for polishing marble).[60]

14 Overlaying with Wax, Polishing with Wax (*Wichsen, Bohnen*)

Primarily wooden household objects with polishing wax. One can also include here the smoothing (*Schlichten*) and glazing (*Glasiren*) of various objects and glove leather with gum (*Gummitragant*), with a mush of fine flour, eggs, milk. Whitewashing.

15 Varnishing (*Lackiren*)

With polishing vanish of amber, copal (*Kopal, Kopalgummi*), and other resinous substances.

16 Tinning, Silvering, Gilding (*Verzinnen, Versilbern, Vergolden*)

17 Glazing, Glazes (*Glasiren, Glasüren*)

Whereby things attain a thin, smooth coating of glass. Sometimes it is advantageous, to quickly give a glaze to crucibles (*Tiegeln*) and other vessels, in which

59 Saffians refers to leather of goat or sheepskin, dyed brightly.

60 Arsène Nicolas Baillet du Belloy (1764–1845) was a French engineer. The original French title is *Observations sur les machines à polir le marbre*.

metals should be melted, by rubbing them or coating them strongly with borax, which glazes.

18 Melting on (*Anschmelzen*)

Sticks of wax are made to shine (*geglänzet*) on coals.

19 Smearing

With tallow, suet, soap, oil, tar. Some drawers, boxes, and casings (*Ueberzüge*), which are somewhat difficult to open, are improved if one makes them smoother with rubbing.

Presses and other machines can, instead of oil, be coated with water lead (*Wasserblei*),[61] which Robert Moray[62] made known to the London Society in 1662. See Birch's *History of the Royal Society*, vol. 1, page 167.

> But even if the old masters have discovered everything, one thing will be always new: the application of scientific study and classification of the discoveries made by others.
>
> SENECA, *Letter* 64[63]

61 Molybdena.

62 Robert Moray (*c.*1605–1673) was a Scottish statesman and natural philosopher.

63 Seneca, *Ad Lucilium. Epistulae Morales*, trans.Richard M. Gummere, Ph.D., vol. 1. (William Heinemann; G. P. Putnam's Sons, 1917), 64. *Sed etiam si ista omnia a veteribus inventa sunt; hoc semper novum erit, usus et inventorum ab aliis scientia et dispositio.*

TRANSLATION

Preface to Johann Poppe's *History of Technology*, 1st Edition (1807)

A half year before my departure from Göttingen I received the flattering offer to write the *history of technology* for the history of the arts and sciences.[1] I resolved to do it with pleasure, because for a long time it had been a favorite idea of mine, to write a history of the handcrafts, manufactures, and fabricks at some point, to which end I already had many useful notes lying in my desk. I thus devoted the last half year of my stay in Göttingen entirely to the purpose of gathering materials for my history with help of the excellent university library, and I even made a few additional journeys from Frankfurt to my dear hometown,[2] primarily with this purpose in mind. I have therefore been fortunate enough to present to the public this Easter[3] the *first volume* of my history.

That under *history of technology* I understand the history of all arts (in the broadest sense) for the refinement of nature products, to which end several sciences have so actively participated, needs hardly to be mentioned. Although the title of this work dictated certain limits as I proceeded, I could not keep from frequently crossing somewhat over these limits by speaking of older inventions, and of inventions from the first years of the nineteenth century. Often I had to briefly mention older history in order to connect the thread of my narrative somewhere, and in the past few years various useful things have been created which it would have been impossible for me to pass by in silence.

I am dividing my entire work into *five main sections*, of which the first contains *the general introduction to technology*; the second contains *mechanical preparations* (the preparations, for which mechanical means are primarily used); the third contains the *mechanical-chemical preparations* (for which primarily more mechanical than chemical means are used); the fourth contains the *chemical-mechanical preparations* (for whose procurement more chemical than mechanical means are necessary); and the fifth contains the *chemical*

1 Found in *History of the Arts and Sciences Since their Restoration through the End of the Eighteenth Century*. Executed by a Society of Learned Men, Section 8, History of the Natural Sciences, Part 4, History of Technology. D. Johann Heinrich Moritz Poppe, vol. 1 (Johann Friedrich Röwer, 1807).

2 Poppe was born in Göttingen, in 1776.

3 For the *Messe*, or book fair.

 | DOI:10.1163/9789004549548_014

preparations (whereby one uses chemical means exclusively). I have organized the *subdivisions* into *paragraphs* according to the various human needs. This way of dividing seemed to me to be the most convenient because it causes the least separation of related objects. One sees this already in the first volume on the history of weaving and knitting. That I, in some secondary areas, could not strictly follow these divisions, that I for example in the preparation of oil had to consider some chemical means used to purify the oil and, for the processing of woolen fabric for clothing also had to consider carpet weaving, was clearly unavoidable. It is, however, impossible to organize the objects of technology into a system in a completely satisfactory way.

As much as possible, no object of technology will escape my attention. But some of them I will only discuss briefly and simply relegate them to an appendix at the end of my work. This appendix should contain both the *corrections* and *additions* which have come to my attention after printing. A complete *register*, however, and an index of all *printing errors* will be found at the end of the publication.

As little as I take – and can take – my work for a finished effort, so very much do I still hope that, for the enrichment of the various branches of technology, it will still contribute its little penny[4] and not remain entirely without approval.

Frankfurt am Main
March 1807.

---

4 "sein Schärflein mit beytragen": *Schärflein* is a diminutive from *Schärf, Scherf*: the least valuable of coins at that time. The idiomatic expression relates back to Martin Luther's translation of Mark chapter 12, verse 42, "sein Scherflein zu etwas beitragen."

TRANSLATION

General Introduction to Johann Poppe's *History of Technology*, 1st Edition (1807)

§1[1] Our planet bears so many beautiful products on its surface that even people who were entirely unfamiliar with the works of creation, and found themselves, as it were, cast upon the as-of-yet-uncultivated orb, would have no difficulty finding sustenance. They would certainly not need to struggle long with hunger or thirst, because they would soon discover something to conquer it. Heat and cold would certainly soon lead them, as rational creatures, to means of protection. Against wind and weather they would know how to look for a shelter. At most, wild animals would cause them discomfort, but soon they would stand as conquerors against these as well. Slaughtered beasts would have to offer them ever new material for satisfying their needs. The flesh could give them sustenance, the hide could give them clothing.

§2 This image, borrowed from raw nature, certainly fits the first humans on earth. One finds in it humans depicted in the condition where they still lived without culture. Means of satisfying the most pressing needs were there immediately after the creation of the earth, but these means were very simple and rough. Could rational creatures really remain standing on this lower level of culture? Certainly not. Rather, it would occur to them *to further refine products of the earth* (*Naturalien*) in order to satisfy their needs in a more complete way. What follows will show the progress humans have made in this regard.

§3 The forces of the spirit, which unite in mankind so beautifully with the forces of the bodies, allow them to sense how many means are still lacking in order to make life much more agreeable. The spirit of invention was active in them to the most beneficial degree. And this spirit of invention soon had to lead them to *process* the products of nature further in order to make their application more useful and comfortable. Gradually, human food was selected and prepared better. The coverings of their bodies received a more comfortable, purposeful form, one built huts and protection against wind and weather, and even made containers for the preservation of food and drink. Bare hands

1 Section One of Johann Poppe's *History of Technology*.

 | DOI:10.1163/9789004549548_015

were clearly no longer adequate for accomplishing the processing of such things. One had to have various pieces of equipment (*Geräthe*) for this purpose, for example, hammer, axe, knife, and things of that sort, which were initially very shapeless and incomplete, but gradually took on a more purposeful and pleasing shape.

Thus man gradually progressed further and further. Once the means were found of better satisfying the most urgent needs, one could think further about new inventions for ever greater comfort. These wishes were also gradually fulfilled. Man's spirit, striving ever higher, was now in flight and could not be kept back. He grasped for new inventions and even strove for beauty in various finished works, and he succeeded. But did man have enough at this point? Did he no longer strive towards new inventions? Did he place limits on the current of his ideas forever? Not in the least. Whatever man took for beautiful and appropriate today, he found the next day to be ugly and unfitting, and thus *luxus* and *fashion* gradually played a part in all his affairs, which were brought to completion by head and hand.

§4 It would be very interesting if one could definitively convey the first occasions and means by which all known inventions for the refinement of nature objects came to be – if one, for example, could say with certainly, how one came upon the idea of breaking down grains of wheat into various parts with ever more perfect mechanical forces, and of separating the pure from the impure. Or, how one gradually came to the idea of applying chemical forces, and used them in part by themselves, and in part in connection with mechanical forces. Or how one, through fermentation, learned how to extract the finest parts from grain and other fruits and, furthermore, learned to spin and weave the hair or wool of animals, and even plant fibers (flax, hemp) and seed-wool (cotton) for pieces of clothing. Or, how chemical means of making animal skins more durable, stronger, and softer, and of preparing them for further processing were discovered. Or how one gradually learned how to separate metal excavated from under the surface of the earth from foreign non-useful parts, to forge it and learn how to make it suitable for various applications, how one gradually invented the means of welding, and invented the bellows and all other tools for it. If one took all of these various inventions together, or even a portion of them, then one must really succumb to astonishment about human talents. One must also admit, however, that a new invention was not necessary for every single process, but it is rather the case that frequently an invention already available was used for a similar purpose, perhaps with a few changes. How many works are there not, for example, that make something smaller, that make something thinner, that make something thicker, that make something smooth or rough, and why could it not be possible, that for one of these various works, which have a certain purpose, the same expedient also serves, even if the materials themselves are not the same? The certainty of this statement is easily proven with ample evidence.

§5 Since the processing of the products of the earth is a labor of the hands, whereby the spirit must certainly always be active, thus one has named it a *handcraft* (*Handwerk*). In the oldest times, every person was a *handcraftsman* (*Handwerker*), because whoever needed something applied his own hand towards its completion. Naturally, women (*Frauenzimmer*) had to preoccupy themselves with handcrafts the most, because wars with men and beasts often called men away from the peaceful herd. And this finally reached the point that one soon left the women to do the handcrafts completely on their own, and the men in their native land went around with axe and hammer at most. Even the most noble women were not excluded from those handcrafts. Thus one says, for example, with regard to the Emperor *Augustus*, that he customarily wore no other clothing than that which he received from the workshop of his sister and daughters. Only among *Jews* did men, even well-regarded civil servants and scholars, preoccupy themselves with various handcrafts.

Weaving, embroidery, and tailoring were primarily the handcrafts which females (*Weiber*) of times past practiced individually and alone. That embroidery and part of tailoring have remained in the hands of women (*Damen*) up through the present day is something everybody knows and is not to be remarked upon. That, however, in some regions of Germany weaving is also practiced by women, for example, in some areas of Lausitz, Mark Brandenburg, Mark Hannover, and Mark Pommern, really deserves our attention and a serious look back in older times.

§6 Handcrafts certainly had to be esteemed as long as they were mastered and practiced by women (*Frauenzimmer*). Once the *Romans* and *Greeks* began, soon after, to transfer handcrafts to their servants and bondmen, they then fell more and more in esteem, and the wives and daughters in all lands slowly distanced themselves from them.[2]

Smiths, swordsmiths, gold- and silversmiths, lathe turners, cartwrights, soap makers, brewers, bakers, cobblers, and other handcraftsmen were comprised

2 1) Tacitus, de Mor(ibus) Germ(aniae) (On the customs of the German peoples), chapters 16 and 17.

of bondmen for whom individual *workshops* were built, primarily in imperial villas and near the cloisters.[3]

They usually stood under the supervision of the mistress of the house. Only seldom did freeborn persons or nobles attend to handcrafts. However, we have examples that occasionally they too completed artisanal projects, particularly of metal.[4] Of the handcrafts which they learned, the smithcraft is named among others.[5] In the eastern lands the monks have to sustain themselves with their own handmade works,[6] and even in the western lands they began to preoccupy themselves with all sorts of handcrafts after the rule of St. Benedict and other laws (*Ordensregeln*) had assigned them particular labors. One surely has to thank the perfection of several useful trades for this circumstance. A few monks achieved so much in various arts, that they could teach their bondmen themselves with great advantage.

§7 The art of painting, the art of sculpture, lapidary art, and other so-called *beautiful* or *free arts* have always been honored, even when the actual handcrafts fell in estimation. But during the revival of the sciences in the twelfth century, these too were seen in a new light. In Germany and in other realms, the number of bondmen became ever fewer due to the Christian religion. The number of buildings increased noticeably, and from monastic workshops, commercial buildings, and peasant huts there gradually emerged villages and finally towns 6).[7] One stretch of land after the other was cleared, and the amount of nature products (*Naturprodukte*) kept increasing. All of this spurred the desire for new and more comfortable establishments inordinately and encouraged the inventive industriousness of the inhabitants.

The consequences of this were, however, also uncommonly beneficial. Not in the year 925 through the emperor *Heinrich the First*, as many believe 7),[8]

---

3 2) (Christian Gottlieb) Riccius, *spicil(egium) jur(is) Germ(anici)* [Gathering of German laws] (Wilhelm Schmidt, 1750). Book one, section 9, page 325.

4 3) (Johann Christoph) Cleffel. *Antiquitates Germanorum potissimum septentrionalium selectae* (Selected German antiquities, especially of the north), Book 10, paragraph 15.

5 4) Muratori, *dissertation inauguralis* (Inaugural dissertation) 45. *Ant(iquitates) ital(icæ medii ævi)* Italian antiquities of the middle ages, vol. 5, column 400. *A. D. Alteserrae. Asceticon seu originum rei Monasticum* [Asceticon or the origin of monasticism]. Paris: Ludwig Billaine, 1674. Book 9, chapter 6, page 465f.

6 5) Ibid. Chapter 7, p. 252f.

7 6) Compare *Spittler, de origine et incrementis urbium Germaniae* [On the origin and growth of German cities]. in *Commentaries of the Göttingen Academy of Sciences*. 1788. Vol. 9.

8 7) Ludewig, *Dissertatione inaugurali differentias iuris Romani & Germanici in opifice exule in pagis* [Inaugural dissertation on the difference in Roman and German laws for handcraftsmen]. Part 1, chapter 3, paragraph 4. in *Collected Disputations*, vol. 2, number 12.

but only in the latter half of the eleventh century was the arrangement made that the inhabitants of castles and towns, who were called burghers (*Bürger*) were allowed to conduct all manner of handcrafts, trade, and sciences. It is very likely we have the club-law (*Faustrecht*) to thank for this arrangement, which thus allowed several other institutions to exist.

§8 Only now, from the handcrafts, did an individual profession of free men emerge who no longer had to be ashamed at all. Already at that time, several handcraftsmen in Germany divided themselves into certain societies, into the so-called *guilds* (*Zünfte, Gilden*) or corporations (*Innungen*), whose establishment rests in part on Roman principles.[9] For presumably they had been transplanted to Germany by bishops from Italy. The main purpose of these guilds was to bring a stricter order to the trade and to keep perfecting the handcrafts themselves. Guild law thus enabled the handcraftsmen to be more respected and gave him honor and status above many other inhabitants of the towns. The resident members of the guilds received the title *master*, whereby otherwise one was only able to use it in the free arts. Only the masters were allowed to pursue their handcrafts on their own and to teach others. They, in turn, had their own superiors, the so-called *overmasters* or *aldermen* and the *guild masters*.

One considered the *certificates of guild membership* or written laws, whereby the guilds were confirmed, to be something very sacred. Every guild had, in addition, its own seal, its own customs, guild houses, hostels, and places of gathering, as well as a unique guild store and counting house (*Kasse*). The guild could itself execute an individual punishment among its members. It could, for example, expel someone from the guild who had committed a significant crime. Whoever did not belong to the guild could neither sell at the market, nor make use of the public work house. Whoever wished to become master had to have learned the handcraft according to law and have worked several

9 The distinctions between *Innungen*, *Zünfte*, and *Gilden* are a bit blurry and the translations here are mostly a matter of convenience. According to one nineteenth century writer, the general name for corporations formed by those who pursue various trades is *Innung*, which can also be called a *Gewerk* when the trade in question is an art or handcraft. *Gild*, by contrast, is a low-German expression whose name comes from the Anglo-Saxon *gild* (contribution, tribute). The name therefore comes from the monetary contribution that individual members made to support the organization. The word *Zunft* comes from the Middle High German *zëmen*, to agree or come together, and was originally more common in northern Germany. See Johann August Eberhard, *Synonymisches Handwörterbuch Der Deutschen Sprache* [Synonymous pocket dictionary of the German language], 13th edition (Leipzig: Grieben 1882) 547–8.

years as *apprentice* and as *journeyman* beneath the master. The apprentice had, in order to learn a handcraft with a master, first prove his freedom and honesty with a *birth letter*, and then become bound to a master. After the completed apprenticeship years, he was *absolved* through the *apprenticeship letter* or declared a *journeyman*. The journeyman, however, had to have traveled or *wandered* for several years and then had to have proven his knowledge through a *masterpiece* before he could proceed to mastership.

§9 Not all handcraftsmen were placed immediately in a guild at that time, and guilds were not established and confirmed in all places at one and the same time. Thus, for example, in Augsburg in the year 1157, according to a town ordinance by emperor Friedrich the First, only bakers, brewers, and butchers were declared associated with a guild.[10] Most of the high country (*Oberländischen*) towns did not conduct much meaningful trade at that time, and thus there were only those handcrafts which took care of the most necessary requirements for sustenance. In *Magdeburg*, there first appeared in the year 1158 a guild of cloth shearers,[11] which was confirmed by Archbishop Wichmann and received many distinctions of honor.[12] At the same location, the same archbishop also strengthened the cobbler guild. At the same time, he allowed it to choose an overmaster within it, who was exclusively authorized to exercise the legal judgements (*Gerichtsarbeit*) for the guild. At the same location, the Archbishop Ludolph erected the painters' guild (*Schilderinnung*).[13] And thus arose ever more guilds here and there. The handcraft ordinances (*Handwerksordnungen*) of Trier, Goslar, Würzburg, Braunschweig, Frankenberg, Wittstock and others occur partially in the twelfth and partially in the thirteenth century.[14]

§10 Germany's handcraftsmen were at that time extremely famous, and most handcrafts flowed from Germany to other lands. Arnold, the father of

10 8) Paul von Stetten, the younger. *Kunst- Gewerb- und Handwerksgeschichte der Stadt Augsburg* [History of arts, trades, and handcrafts of the city of Augsburg]. Augsburg: Conrad Heinrich Stage, 1779. Octavo. Introduction, page 3.

11 9) "Chronica Magdeburgense ad annum 1158" [Chronicals of Magdeburg through the year 1158], in Heinrich Meibom, *rerum germanicarum* (on things German), vol. 2, page 329.

12 See Ludewig, vol. 2, p. 389. Compare with Siebenkees *Beyträgen zum deutschen Rechte* [Contributions to German law], part IV, page 230f. And J. F. Runde. *Grundsätze des gemeinen deutschen Privatrechts* [Principles of the general German civil law]. Göttingen 1801. Octavo, page 389.

13 11) *Heinrich Meibom*. Script. rer. Germ. a.a.O. See Adelung (final edition), s.v. "Schilder," where he defines it as a somewhat unusual term for a painter.

14 12) Johann Carl Heinrich Dreyer, Dissertatio de privilegiis opificum [Dissertation on the privileges of handcraftsmen]. Köln 1751, p. 7.

Count Baldwin III of Flanders, summoned many German handcraftsmen and artisans into his states and granted them great freedoms, and thereby laid the foundation for the famous Netherlandish fabric manufactures, among others.[15] From the year 1186 onward, Germans also developed many useful arts in Liefland.[16] And Italian monks (*Umiliati*)[17] learned wool weaving among others things from the Germans, and shared it with their countrymen who, supported by excellent artisanal talent, already knew how to thrive.

The Germans, especially the lower Germans (*Niederteutschen*), distinguished themselves in weaving, primarily from the twelfth century onward.[18] But they also made much progress in all sorts of metal-, wood-, ivory-, stone-, and glasswork. One still sees this in many things which are found here and there in curiosity cabinets and among church treasures. And which excellent vessels of valuable stones, which superb harnesses of gold, completed by Germans, did Emperor Otto the Great not give to the British king Adelstan at the marriage of his daughter? What valuable cloth and garments from leather, which Germans had also made, did not various foreign princes not wear as gifts from German princes? In short, Germany had at that time the greatest masters in almost every handcraft. And furthermore, as the continuation of my work will show, do not also most inventions, whereby handcrafts and arts are perfected, and many new extremely important trades are developed, come from Germans? Here, I will only mention as an example the art of book printing, the art of etching and engraving, oil painting, the preparation of gunpowder, and watchmaking, which have only Germans to thank for their origins.

§11 The cessation of servitude among handworkers (§ 8) had the most beneficial influence on all kinds of wares and artisanal works. It was impossible for people to work with willingness who lived in a state of such low esteem, who had to perform compulsory labor, and who had to work completely uncompensated for the emperor and the sovereign officials whenever they traveled through their location. They received more incentive and desire to work when they were freed from this oppressive state. And thus it was no wonder that the products which came from their hands turned out much better, and that the

15 Adam Anderson's *Geschichte des Handels* [History of trade]. Translated from English. Part 1, page 406f.

16 14) Johann Daniel Gruber. *Liefländische Chronik* [Livonian chronicles]. Part 1, page 7.

17 Humiliati, an order thought to have been founded in the 12th century, suppressed in the 16th century by a papal bull.

18 15) Michael Ignaz Schmidt. *Geschichte der Deutschen* [History of the Germans]. Vol. 2, p. 98.

handcraftsmen endeavored to find an ever more perfect refinement of the raw products.[19] Certainly, they retained various signs of abasement for a time. But they also soon worked themselves upwards from this state, with courage. Thus, for example, the emperor could, during his stay in the imperial towns, marry the daughters of handworkers there to his household servants through a mere proclamation. The handcraftsmen of the imperial towns also gradually freed themselves from this compulsory power. The handcraftsmen in *Frankfurt am Main*, *Wetzlar*, *Friedberg* and *Geinhausen* were the first whom this benefited by the privileges granted to them by the *Emperor Heinrich the Seventh* and *Emperor Richard*.[20]

§12 The expansion and improvement of the existence of the handcraftsmen in the *thirteenth century* both within and outside of Germany was very quick and beneficial. The king *Louis the Eleventh*[21] of France, who in the year 1270 shortly before his second crusade made great use of the establishments of St. Louis, had the handcraftsmen organize themselves into guilds at the same time, through the help of Étienne Boileau,[22] and he gave them wise statutes that soon afterwards were taken up and imitated within all communes of that time and thereafter.[23] But as it goes, fortune and the desired continuation of a business often make the people that pursue it proud and overconfident, and in others, though it is not their business, it commonly produces hatred, envy and greed. So it happened to the handworkers and arts in France, and so it happened to them in Germany and in other lands. Many persecutions and hardships rained down upon the handworkers. The emperors of that time, particularly *Emperor Friedrich the First* (Frederick Barbarossa) (1122–1190) and *Emperor Friedrich the Second* (1194–1250), sought to bring down again all handcraft collectives (*Handwerksverbindungen*) that had scarcely begun, and in many places they even disbanded the trade organizations. For example, *Emperor Friedrich the Second* tore down the trade organizations in Goslar in the year 1219.[24] His successor, *Rudolph of Habsburg* (1218–1291), on the other

19 16) *de conditione et statu servorum* in *Germania* [On the condition and status of slaves in Germany], Joachim Pottgiesser. Book 3, chapter 1.

20 17) Franz Domenicus Häberlin. *Entwurf einer pragmatischen deutschen Reichshistorie* [Draft of a pragmatic history of the German empire], vol. 2, pages 461f.

21 It was actually Louis IX.

22 Étienne Boileau was appointed provost in 1261 and wrote statues of regulations on matters of police, trade, and industry.

23 18) Jean Joinville. *Histoire de St. Louis* [History of St. Louis] part 3, page 124.

24 19) Johann Gottlieb Heineccius. Antiquitates Germanicae. [German antiquities]. Book 3, page 301.

hand, united them again, and also reinstated the disbanded guilds in other locations.[25]

A main reason for the disbanding of the guilds was, however, the great importance which the handcraftsmen in the towns gradually acquired and which even drove them to attempt to remove themselves from the governance of the territorial lords. For that reason, Bishop *Heinrich the Second of Worms*, in the year 1234, also forbade all handcraft collectives, and did all possible harm to the members of the guild in the sales of their wares.[26] In a similar way, the earl *Heinrich the Second* of Hessen, in the year 1368, disbanded the guilds of *Frankenberg*, with the single exception of the wool weavers.[27] The same thing occurred, as we will soon (§ 15, 16) hear, in several other lands.

§13 All efforts of the princes to suppress the guilds did not, however, last for long. The status and regard for handcraftsmen did not return to the former low level. Rather, they were elevated ever higher. Nobles began to get involved with the condition of commoners and to marry them. The ambition of the handworkers received increasingly more nourishment, there was no end at all to the striving for greater respect, and this finally went so far that the handcraftsmen not only demanded to take part in the governance of the city, but even to rule it on their own. Thus in several cities, namely *Würzburg*, *Goslar*, *Braunschweig*, and *Lübeck*, one came into such a confrontation that resulted in bloodshed.[28] The imperial estates (*Reichsstände*) took it upon themselves to retain, from old freeborn and monied families (*Geschlechtern*), the exclusive possession of important governmental town offices, and to push back the handcraftsmen into the proper confines of dependency. All guilds and other societies of handcraftsmen were to a certain degree subjected to the town authorities. No assemblies and gatherings were allowed to take place without the permission of the magistrate, or without a person of the council in attendance. Also, the magistrate was now permitted to weigh in with regard to punishments.[29]

---

25 20) Johann Christian Lünig. *Reichsarchiv* [Imperial archive]. Part 1, p. 858.

26 21) Chronica Wormatia [Chronicles of Worms], Ludewig. in *Reliquiae Manuscriptorum* [Remnants of manuscripts], volume 2, part 3.

27 22) Chronica Frankenbergense [Frankenberg chronicles]. Justus Kuchenbeker. Annal. Haff. Coll. v. 102.

28 23) Carl Renaus Hausen, *Staats-Materialien und historisch-politische Aufklärungen für das Publikum* [State materials and historical-political explanations for the public]. Piece 3, page 279. Poppe's *History of Technology*, vol. 1.

29 24) Christian Gottlob Haltaus, in *glossarium germanicum medii aevi* [German glossary of the Middle Ages], columns 1028f.

Furthermore, from the side of the authority, the correction of many faults of the artisanal system (*Handwerkswesen*) was also not forgotten. Many ordinances and privileges, particularly the Wittstockian ordinance from 1278, did not have the intention of punishing the handcrafts at all. One wanted, for example, not to allow handworkers in the countryside. In the towns, however, various buildings were erected for the more easy preservation of the handcraft's stock, which received the name work houses, and whose use one finds preserved up to the present day. *Guild letters* or written laws (§ 8) were still introduced by many guilds which previously had none.

§14 In the Netherlands, in particular, the handcrafts achieved a glorious bloom at this time.[30] The people of the Netherlands had the excellent opportunity for purchasing raw materials and selling finished wares. This in turn had to increase activity more and more. The greatest distinction therein came in part to the people of Flanders. But that also incited the magistrates in this earldom to impose strict levies on the guilds. The handcraftsmen, among whom the wool weavers played an especially important role, rose up in strong protest against this. In the beginning, they left the towns and moved to the nearby countryside. Since, however, they were also oppressed by the same sanctions from this location, they left the lands around Flanders entirely and moved over to Brabant in great number. Here they were received with open arms, and this good reception subsequently attracted more and more talented workers. *Löwen* alone was supposed to have received in this way around 4,000 master weavers and 15,000 journeymen.

The quiet and peace of the handcraftsmen did not last for long in Brabant, however. Soon, they suffered all kinds of hard persecutions here as well. There occurred many episodes of unrest, and even bloody scenes, whereby the handcraftsmen were forced to seek their wellbeing in foreign lands. Most of them emigrated to England. On this fortunate island, they practiced their trade, primarily wool weaving and the preparation of cloth, with great freedom and thereby laid the basis for the excellent manufactures which consequently gave England such a great preponderance over all other lands.[31] Already by the beginning of the twelfth century, England had rudimentary cloth-weaver

---

30 25) Historical times judicial analysis with general comments, serving to answer the question: What was the state of the handicrafts and of the commerce in the Netherlands, at the time of the thirteenth and fourteenth centuries? Who received the prize from the Imperial and Royal Academy of Brussels in the year 1777 by Mr. W. F. Verhoeven; in *Memoires on questions asked by the Imperial and Royal academy of humanities in Bruxelles*, 1778. Quarto.

31 26) Adam Anderson's *Geschichte der Handlung* [History of trade], part 2, page 221.

guilds. But only towards the end of the thirteenth did the fabrication of its cloths have its beginning, once *Edward the Third* allowed a generous residency in his country for those cloth-weavers.[32] A portion of the wool workers had also now emigrated to the northern part of the Netherlands, namely to Holland, Seeland, Westfriesland, and Obernssel. These states were also elevated thereby. It is easy to imagine that the envy of the residents of Flanders erupted into flames because of this. These flames were suppressed again. In the year 1304 it really came to a sea battle between the inhabitants of Holland and Flanders, whereby the former were the victors.[33]

Germans who emigrated to England built their own *guild house* in London, from which location their wares poured out of London into every part of England. King *Henry the Third* gave this German guild house considerable privileges in the year 1257, which King *Edward I* again confirmed twenty-three years later.[34] Many German handcraftsmen were already summoned in the 13th century by the Norwegian kings to the famous trading town of *Bergen*, where they received many rights and freedoms and helped develop several beneficial kinds of sustenance, above all metalworking and leatherworking 30).[35]

§15 Throughout the entire fourteenth century, there was no end to the fighting between handcraftsmen and authorities. In the years 1303 and 1352 the handcraftsmen in Augsburg tried very hard, in vain, to introduce the guild regiment. Only in the year 1368 did they succeed after strenuous efforts. In Vienna, in the year 1364, Prince *Rudolph the Fourth* of Austria once again disjoined all trade societies and other groups of handcraftsmen and annihilated all decrees and ordinances previously held by the handcraftsmen. Landgrave *Heinrich the First* of Hessen confirmed, in the year 1307, all trade societies in *Cassel*. Whoever had acquired guild status there could also practice his handcraft. Prince *Otto of Braunschweig* also allowed similar ordinances to be made in his lands, which initially applied only to the cloth-shearers. We already know that

---

32 27) Thomas Madox, *firma Burgi*: or, a *Historical Essay Concerning the Cities, Towns, and Buroughs of England*. Chapter 10, section 20.

33 28) Ludovico Guicciardini, *omnium Belgii sive inferioris germaniae regionum descriptio* [All Belgium or regional description of lower German lands], p. 30.

34 29) Franz Domenicus Häberlin, *analecta medii aevi* [Excerpts of the Middle Ages], page 7, number 2; page 9, number 3. Georg Sartorius, *Geschichte des hanseatischen Bundes* [History of the Hanseatic league]. Part 1. Göttingen. 1802. 8. Page 283.

35 30) Ludvig (sic) von Holberg. *Beschreibung der berühmten Haupt- und Handelsstadt Bergen in Norwegen* [Description of the famous capital and trading town, Bergen, in Norway]. Part II. Kopenhagen und Leipzig 1753, page 26.

Heinrich the Second of Hessen dissolved the guilds in *Frankenberg* once again (from § 12).

§16 Just as all the guilds of the handcraftsmen were perpetually unsettled, a few really collapsed completely, but this did not mean that they were destroyed forever. The faltering ones were kept, the destroyed ones erected again. They received honor and respect once more, their own statutes and ordinances, and even aristocratic members, in particular leaders and heads of craft guilds. Every handcraft even created for itself a complete coat of arms,[36] and scholars and civil servants became guild brothers. The flourishing of the handcrafts, and with them trade, was thereby encouraged immensely. The exaggerated measures of the guilds kept going, however. At the end of the fourteenth century, in *Braunschweig*, *Lübeck*, *Wismar*, *Rostock*, and *Hamburg*, they caused great unrest against the magistrate who thereby lost his regiment for several years and was even expelled from the towns. Only in the year 1416 could the old form of government be reinstated.[37]

In the fourteenth century, many inventions and discoveries became known which had an uncommonly important influence on the artisanal system. From the *preparation of gunpowder*, which one had certainly known about long before the supposed year of its invention in 1354, there arose in this century a new trade which soon afterwards generated a new handcraft, the gun foundry (*Stückgießerey*) and production of weapons. The invention of *rag paper* probably already occurred in the thirteenth century. In the fourteenth, however, the preparation of this paper really took off. The king *Roger* of Sicily had, in the year 1148, encouraged silk weaving by Greek workers to take root in Italy. In the year 1314, after the destruction of *Lucca*, some of the Greek silk weavers went to Germany and practiced their handcraft there. All kinds of weaving, of gold, silver, and other metal works gained ever greater perfection in the fourteenth century.

§17 The handcrafts and arts naturally flourished the most in Germany, England, French, Italy, and in the Netherlands. Every one of these countries had its own advantages in the processing of these or those natural products.

---

36 31) See, for example, Heinrich Nettelbladt, *Historisch-diplomatische Abhandlung von dem Ursprung der Stadt Rostock*, page 88. On the harness of the handcrafts and the membership dues (*Harnischgeldern*) of the guilds.

37 32) Compare, among others, Carl Ribbentrop, *Vollständige Geschichte und* Beschreibung *der Stadt Braunschweig* [Complete history and description of the town of Braunschweig]. Part 1, page 67f; Johann Rudolf Becker, *Umständliche Geschichte der freien Stadt Lübeck* [Thorough description of the free town of Lübeck]. Part 1, p. 320f.

But also other countries, for example Portugal, Spain, Denmark, and Sweden, made notable progress in the refinement of them. Due to the crusades, many countries had naturally lost various useful handcraftsmen. But several useful arts were brought back by crusaders into these same countries.

Europe in general has, for many arts and handcrafts, foreign parts of the world to thank, whose inhabitants were initiated in them much earlier. How many come not from the Orient? Do not the Egyptians have the first weavers, the Phoenicians the first dyers, and the Arabs the first distillers, also of spirits?

§18 In the fifteenth century, the artisanal system did not go through any significant change with regard to its external form and constitution. In a few locations, handcraftsmen succeeded in elevating themselves to councillorships, and pushing through their own plans for their alliances. In other places, the territorial lords dissolved the old constitutions for crafts (*Handwerksverfassungen*) and gave them new ones, wherein obedience to authority was required. Thus, for example, the landgrave Friedrich von Thuringia, in the year 1414, destroyed all of the craftsmen's guilds in Chemnitz and proclaimed new ones under the supervision of the magistrate. These new guilds were not permitted to draft articles for the constitution of their organization (*Innungsartikel*) except with the help of the town magistrate.

The guilds of some places only attained a director in the fifteenth century, for example *Stuttgart*. In such places, the handcrafts must have matured to more completion later than in other places. The German Hansa had, among their laws of governance (*Polizeygesetzen*), primarily taken the state of the handcrafts into consideration. A guild member (*Zunftgenosse*) who left one hanseatic city and wanted to ply his trade in another had to first present favorable testimonials from his former magistrate. One would not be at all satisfied with only the affirmations of his guild superiors. No master of a handcraft was allowed to accept guildsmen in his employ without providing their names to the directors and guild masters. These, in turn, were not allowed to allocate any monetary fines to guildsmen, or to strike their name from the guild register, without the magistrate's prior approval. No new article in the guild's constitution would be declared valid without the approval of the magistrate.

§19 The fifteenth century was also rich in various kinds of inventions and improvements. Shortly before the middle of the century, the *art of book printing* had its beginning. Already through this art alone the Germans, as inventors, have won the greatest renown before all other nations. With the art of book printing comes the sibling art, *type-foundry* (*Schriftgießerey*), which emerged at the same time. And also the *art of wood carving* (*Holzschneiderkunst*). A *craft of book binding* did not exist before then. This too owes its existence merely to the art of book printing. Only a few years later, Germans invented the *art of copper plate engraving* (*Kupferstecherkunst*) and the *art of etching* (*Aetzkunst*). A German invented in the year 1472 the pedal organ, and Germans brought already in the year 1410 the old German art of *painting with oil colors* to greater perfection and applied it to wallpaper, among other things. It would be impossible for me, in the *introduction*, to touch upon all inventions and improvements which came to light due to Germans. The further continuation of my history will provide more detailed information on this matter.

§20 Even the foreigners, however afire with envy and jealousy they may have been, had to give German inventiveness and skillfulness its due. Their reputation in all kinds of works, particularly in *mechanical* works, spread to even the most distant peoples. The Italian bishop *Paul Jovio* confirmed that his countrymen, well into the fifteenth century, had received the best artists, *architects, painters, sculptors, stone engravers* (*Steinschneider*), *copper-plate engravers, mechanics*, etc., from Germany.[38] Even the Italian *Niccolò Machiavelli* stated the same thing. He admits that Germany enriched all of Italy with many useful inventions and provided it with many of the most skillful craftsmen.[39] The Cardinal *Aeneas Sylvius Piccolomini* admitted that the Germans of his time were superior in architecture before all other people.[40] Another Italian, *Johann Anton Magino*, praises the Germans as the originators of many ingenious inventions. From them, he says, stems book printing, the art of canning, of watchmaking and metalworking.[41] Yet another Italian, *Johann Boteri*, confirms the same thing and adds that the Germans had at that time even surpassed the Romans in their industry.[42] The Athenian *Laonikus Kalkondylas* claimed that one would not easily find in mechanical works a nature that had accomplished what the German did. The Germans had the greatest advantage in all the arts. Just how famous Germans also were in bell founding, is shown by many German clocks which are found in foreign realms to this day. No less experienced were the Germans in beer brewing, wine making, weaving, and the creation of fabric.

38 33) Elogium Virorum illustr. [Praise of famous men]. Basil. 1578.

39 34) Friedrich Christoph Jonathan Fischer's *Geschichte des teutschen Handels* [History of German trade]. Part 2. Hannover 1785. 8. Page 493.

40 35) Letter 170 to *Joan. Campit*, see Fischer and others.

41 36) *Geographia*, part 2, p. 201.

42 37) In Relat. Europ. P. I. L. I. de German.

§21 No German town found itself at the end of the fifteenth century in such a flourishing state as *Nuremberg*. All reports which have come to us from those times confirm that this town had an excess of artisans and handcraftsmen who acquired fame throughout Europe for their excellent works,[43] as no other place could boast of the same distinction. Nurembergian wares were sent out not only to England, Spain, France, and Italy, but also to the Levant. One even saw in the seventeenth century a repository of Nurembergian mirrors and other trinkets in Cairo, Egypt.[44] In the year 1430, a proper guildhall for cloth was established, which received its own ordinance in 1431. In short, Nuremberg had in the fifteenth century, as in the following sixteenth and seventeenth centuries, many excellent artisans and handcraftsmen, namely sculptors, painters, copper-engravers, model-makers, embossers,[45] stone carvers, organ makers, instrument makers, lathe operators, medal makers, and all sorts of mechanics and other artisanal workers.[46] There were also iron works, smelt works and other similar operations already in the neighborhood of Nuremberg.[47]

In the town of *Augsburg*, whose workers subsequently competed with those of Nuremberg, the handcrafts seem to ripen to greater perfection later, only in the sixteenth century. Weaving mills were already at work there prior to the fifteenth century. Gold-, silver- and metalworkers first appeared, as far as one can know, in the thirteenth century. In the three following centuries, these trades were much refined, but nevertheless they were scarcely in the position to keep pace with the citizens of Nuremberg. Weaving mills first appeared in Augsburg only in the fourteenth century.[48]

§22 In the *sixteenth century*, the handcrafts and arts kept increasing in perfection. The Germans again distinguished themselves the most through new inventions. But many useful discoveries and improvements also spread

43 38) Franciscus Irenicus, *Exegesus historiae Germaniae* [Explanation of the history of Germany], book 3, chapter 112.

44 39) J. Cochleae Commentar. in Pomp. Mel. p. 33.

45 See Adelung (final edition), s.v. "bossiren," where he lists *Poußiren* as a synonym.

46 40) Johann Gabriel Doppelmayr, *Historische Nachrichten von Nurembergischen Mathematicis und Künstlern* [Historical reports on mathematicians and artisans of Nuremberg]. 1730. Folio.

47 41) Conrad Celtes. *De origine, situ, moribus et institutis Norimbergae* [On the origins, site, habits and institutions of Nuremberg]. Chapter 2.

48 42) Paul von Stetten. *Kunst- Gewerb- und Handwerksgeschichte der Stadt Augsburg* [History of arts, trades, and handcrafts of the city of Augsburg]. Augsburg: Conrad Heinrich Stage, 1779. Introduction, page 5.

towards Germany from foreign lands, particularly as concerns the weaving of wool, cotton, and silk.

The inhabitants of the Spanish Netherlands were successful inventors of useful tools and busy improvers of various kinds of work. The English people certainly owe them almost all of their later flourishing in all kinds of handcrafts (particularly in manufactures and fabricks). As the Spanish Inquisition began to work its mischief in the year 1565, an inordinate number of industrious people left the Netherlands. They emigrated mostly to Germany, Sweden, France, and England. The number of wool weavers alone, which departed from Flanders, is thought to be 100,000. They found the most welcome from the English, for whom they made – and taught how to make – so many useful things, and particularly all kinds of beautiful cloth wares. Germany, primarily Saxony and the Mark Brandenburg, received many talented Netherlandish handcraftsmen, who also established the foundation for later manufactures and fabricks.

§23 Only in the particular history of every individual trade can I introduce and precisely examine individual inventions and improvements. Here I speak only of the introduction, expansion, and improvement of handcrafts and arts (whereby I also mean the manufactures and fabricks) and of the advantages which this had for the whole human race, in general.

Nuremberg, that famous Nuremberg, still drew to it many talented workers from Italy, France, and the Netherlands in the sixteenth century. To these belong, among others, the dyers and cloth-dressers (*Tuchbereiter*) from Antwerp, to whom the town council granted various impressive privileges in the year 1569. For that reason, it could, among all German sisters, always triumph more. For that reason, it gained more and more in wealth and regard. But other German towns and lands did not in the least remain *very far* behind Nuremberg. Many places could demonstrate handcrafts and arts that Nuremberg did not have. In Hessen and Bohemia emerged, just past the middle of the sixteenth century, the first *alum works*. The glassmaker *Schürer* from Saxony invented the *color cobalt* or *smalt*, which subsequently played such an important role in the art of making glass and porcelain. The inhabitants of Holland were the first to draw the secret from him. They constructed dyeing mills; *Schürer* himself however founded the first of Saxony's *indigo dyeworks*, which later became so famous and which delivered the material in question in such great quantity and in a high degree of perfection. Glass painting gave birth to the art of *enameling*, which appeared in the Netherlands.

Germans at this time also made the most effort to bring *the art of mining* and the entire field of *metallurgy* to a greater degree of perfection. New buildings, new tools, and machines were invented in order to extract metals from ore deposits in a more comfortable, advantageous way, and thus also to use them. Methods were discovered for saving furnaces and workers, to avoid dangers while working, etc. In short, it is impossible in this *introduction* to list all the advantages which handcraftsmen and artists thought of for the perfection of their trade and used in its practice.

§24 Handcrafts and arts were now really on the rise and could no longer be held back. The *seventeenth* and *eighteenth* centuries were therefore even richer in many inventions and discoveries. The actual *manufactures* and *fabricks* for all kinds of works increased extraordinarily. In particular, England and France gained the upper hand above all other countries.

If *several* handcraftsmen in one or various arts came together for a *single* purpose, namely, to work on the refinement of some kind of raw product, then one gave their trade the more significant name *manufacture* or *fabrick*: *manufacture* if hands were used directly, or in the absence of machines for the completion of the wares and *fabrick* if one used fire and hammers for that purpose. The true distinction between them is often not easy to discern. A maker of fabrics or woolen material, who only does so with the help of a few journeymen (*Gesellen*), pursues the handcraft of making fabric or material; a hatmaker who completes hats with a similar degree of help pursues the art of making hats. If, however, many workers are present in the workshop, of whom one always passes the work along into the hands of another, then the handcraft of making fabric or woolen material, and from the handcraft of hat making, develops into the manufacture of fabric or woolen material and the manufacture of hats. The smith in town is a handcraftsman, but the hammer-works are fabricks, for example: iron fabricks. Thus there are knife smiths and knife fabricants, button makers and button fabricants, watch makers and watch fabricants, etc. Some works cannot be produced except in bulk, for example, porcelain making, glassmaking, etc, one tends therefore to never call these handcrafts, but rather always manufactures or fabricks.

The designation 'manufactures' has only been introduced in recent times, although some of these works, for example weaving in bulk, were already pursued in the thirteenth and fourteenth centuries. The actual manufactures and fabricks arose later, however. One considers the residents of Nuremberg the first among the Germans to construct their own fabricks. No one could mistake the advantages of manufactures and fabricks. Through them, all works were completed more quickly and accurately. The cause of this was not only the greater skill which all workers in the manufactures and fabricks achieved through constant practice, in that one worker only completed this part, another worker only that part of an object, but rather, primarily, through the introduction of *machines* (instead of mere hand tools), or those artificial contraptions whereby works were performed with less force and time, and much more uniformly.

§25 In the middle of the fourteenth century a German invented the *machines for wire drawing* (*Drathziehen*). Another German invented pocket watches in the last years of the fifteenth century, and (most probably) it was again a German who invented, at the beginning of the sixteenth century, the first *sifting machine* (*Beutelwerk*) for grain mills. An Englishman created, at the end of the same century, the first *stocking frame* (*Strumpfwirkerstuhl*). In the seventeenth century, a German invented the first mint works for coins, and in the beginning of the eighteenth century, yet another German invented the so-called *Hollander* for paper mills.[49] And did not an Englishman make himself famous directly after the middle of the eighteenth century through the invention of the *carding and spinning machines* (*Kratz- und Spinnmaschinen*) for the manufacture of cloth? All these and many other inventions in turn led to a great number of new ones and were a primary reason for the extraordinary perfection of the important manufactures and fabricks with which England, France, Germany, the Netherlands, Switzerland and many other lands are so abundantly furnished.

§26 Such manufactures and fabricks, which completed things of significant value or important use, contained ingenious tools and machines. With them, one could facilitate the work and improve the occupation of the spirit. There were also those which processed less important and useful things, even things for amusement, but one should not be any more surprised by this than by the first case. Through the construction of carding and spinning machines, stamping machines (for coins and other metallic wares), weaving machines, machines for cutting cloth and fabric, machines for the knife-smithing of knife and dagger blades, file-cutting machines, mirror and metal grinding machines or polishing machines, cutting machines for clock wheels, and such things, one will naturally deliver better and more inexpensive wares. But why should one not desire the same for the completion of billiard balls or clay balls, the Jew's harp, laces, the candlesnuffer, and such things? All the more so that for the sale of these inexpensive things it depends upon the quantity, if people are to live properly from such a trade. The acknowledged artisanal industry of the people

49 Hollander beaters produce pulp for paper-making.

of Nuremberg gives us the most convincing example of how much is also to be earned through small things, if one knows how to complete them quickly and in great quantity.

§27 The industriousness of the people of Nuremberg, which was perhaps not insignificant even before the fourteenth century, is still always best recognized in the metal works. Since long ago, many common household objects and toys for children had been made in Nuremberg which were sent in great number to all European lands, and which in fact comprised a noticeable branch of trade. The *Jew's harps*, for example, went and still go by the centner (*quintalweise*) to Poland, and even to Crimea, where the nomadic people entice the most beautiful music from them.[50] Candle snuffs, larding needles, etc., are certainly simple but still really useful wares which are used in unbelievable number. As the consumption of coffee in Germany became more common, one roasted it in pans and put it in mortars, as the Turks still do. But soon *coffee drums* and *coffee mills* began to appear, which, even if they were not invented in Nuremberg were still quickly imitated in quantity. Even now they are exported by the million from Nuremberg, since most people in Germany still grind their coffee with Nurembergian coffee mills.

In almost all wares from Nuremberg there certainly is a tendency towards fussiness, an elaborate daintiness which does not appear pleasing to the eye but whose usefulness is in no way to be mistaken. Over a hundred years ago these were things much admired in Germany, so that this fussy taste thereby became ingrained in our fatherland. As of seventy-odd years ago, however, various household wares and the like came from France. These French wares looked better, but were designed for quick enjoyment and consumption. The decorations on them were light and delicate, but excessive, colorful, and grotesque. One considered them tasteful for a while and thus suppressed the wares from Nuremberg. But also the French taste did not last long. The English taste reached us, and the French quickly yielded to it.

The English wares primarily distinguish themselves in that they are worked with understanding to be durable, good, and simple. The English had made good progress in the visual arts. Through them, antique forms soon became common. The mechanical artists and handcraftsmen often made use of them in their work. That which is plain, grand, and purposeful in their work soon went over to Germany, and Germans now undertook to do the same as the English.

---

50 43) *Berlinische Monatsschrift* [Berlin monthly periodical]. Friedrich Gedike and Johann Erich Biester. Vol. 1. Berlin 1783. 8. February, p. 199.

§28 No place in the world can boast of producing wares as cheaply as the people of Nuremberg, and after them those of Augsburg. This cheapness has, however, inordinately advanced the sale of wares. If handcraftsmen are industrious and modest, then their wages can also be less. And the people of Nuremberg had always possessed these virtues to a high degree. *Nuremberg* and *Augsburg* are also, as far as I know, the first places in Germany, and certainly among the first in all of Europe, which processed their wares *in* fabricks (§ 24). That had to have had a beneficial and useful influence on the inexpensiveness. How much is not spared alone through *machines* (§ 24, 25) in time and wages? To these belong, for example, the various kinds of mills on the Pegnitz river for sanding, polishing, cutting, lathe turning, etc., which have long accomplished excellent service.

Even up to the present day, many wares are made in *Nuremberg* so extremely inexpensively, as in no other place in the world. One could take as an example a roll of *brass music wire*. How many devices are necessary to make this wire? How much work does it require, and through how many hands does it go before it has received the required thinness? A roll of brass wire of the finest kind costs only 1 Kreuzer[51] in *Frankfurt*, 2 Kreuzer for the strongest kind, and whoever takes many roles at once gets them for an even better price. And yet the wire found in such a roll is at least two times the length of a piano. If one calculates in addition the lathed roll itself, whereupon the stamp is pressed, and in addition the work of winding it, the transport from *Nuremberg* to *Frankfurt*, the customs, the service of the merchant in Nuremberg and Frankfurt, then one really has to be astonished at the low price. – Equally striking examples of how far the industry has arrived in other locations are, among others, the leather shoelaces fitted on each end with a metal covering (*Hülse*), of which one prepares in *Elberfeld* an inordinate amount, and of which the salesperson in a place sells for only one Heller per piece. Further, there are the wooden wall clocks made in the Black Forest with moving pieces and chimes, of which one piece scarcely costs an imperial thaler (*Reichstaler*). And who does think, in this regard, about the sewing pins and needles and other so-called trivialities (§ 27). If for all these things one had no machines (which usually are driven by water), then the price would without a doubt become much higher, and have to be that way.

§29 The people of Nuremberg have kept some inventions, and even some handcrafts, just for themselves for several centuries. Nuremberg's handcrafts

51 A small coin. There were sixty Kreuzer in a Florin.

belong namely to the so-called *locked* handcrafts, which means that the masters are only allowed to take the sons of town residents into the handcraft, and these must swear an oath not to reveal any step of the work or any invention to a foreigner. As natural and appropriate as it is for someone to try to remain alone in the possession of his own invention, in order to be rewarded for his trouble and effort, it is also harmful to the general wellbeing. For secretive people are usually treated secretively by others. On the other hand, trust is worth the trust of another. *Nuremberg* was in perfect bloom as long as it alone remained in possession of many inventions. Since, however, many things which Nuremberg used to produce are no longer necessary, and since many handcrafts and arts which this town used to possess exclusively for itself, for example gold beating, and the making of gold wire, not only expanded to nearby *Fürth*, but also to distant places like *Berlin*, *Leipzig*, etc., since that point, the decline of *Nuremberg* has really begun.

One of the main characteristics of handcrafts from Nuremberg and also even from the inventions created there is that they are all soon brought to a certain level upon which they remain without further elevation. It sounds strange that a town which is so full of industry does not think of further improvement, and that in all of its works always the same old way of doing things carries on. And yet this is proven. What a contrast, if one compares the iron and brass works of *Birmingham* with those of *Nuremberg*! The English are always thinking up new inventions and improvements which can please the public. But the handcraftsmen and artisans in our otherwise so famous German town of industry do not do the same. Without active help, Nuremberg would remain ever further behind, and finally the most beautiful fruit-bearing branch will pass away.

§30 The first means of helping *Nuremberg* return to its once glorious flourishing would be the removal of the lock among the handcrafts. For there is no use in looking away from it: many locked handcrafts have already been practiced elsewhere for a long while. And what is worse, on the one side the people of Nuremberg lock their handcrafts, and on the other side they turn away talented people from the town for trespassing on the sixth commandment. As wholesome as the purpose of this strict moral doctrine may be, in order to keep the honor of the guilds from harm at any cost, just as invalid was, however, the practice of those laws as neighboring cities began to receive with open arms those industrious immigrants and talented people. The town thereby lost several useful workers who now made a good life in foreign places. In this way, *Berlin*, for example, received in 1586 its first brass beater (*Messingschläger*) from *Nuremberg*, took on many assistants, and made a large business. And thus other cities have also, over many years, received a number of useful citizens from *Nuremberg*.

Attaining the right to be a master craftsman is expensive in Nuremberg, and this certainly hinders the progress of industry. In addition, few houses alone possess the right to be allowed to pursue certain trades – for example, brewing, baking, smithing, etc. – as is also the case in several other towns. Such houses are even forbidden to be sold to any other trade, under penalty of confiscation, even if they remain unusable and empty. Plenty of ways to keep the industry from its best progress!

§31 In *one* country the regard for the handcrafts rose often, and in the other it sank. At *one* time the practitioners of the handcrafts were greatly respected, and in another less so. As one began, in Germany, to permit handcraftsmen some advantages, and even to give them respectable positions (§ 18), there arose the expression *honorable* and *praiseworthy handcrafts*.[52] This designation subsequently remained, but not the honor and respect attached to it. In public life, the position of handcraftsman suffers from some contempt. Those in other trades consider it something common and believe that a handcraft requires very little intellectual ability. But however common such thoughts may be, how off the mark such judgments are! To every handcraft belong methods, proper tool use, precautions, and judgements, which are difficult to figure out for someone who never deemed them worth consideration. How often have thinking persons, with great advantage, made a new and important use of them!

Handcrafts that presuppose great natural abilities and much knowledge have up to the present day retained the name *arts*, for example, the art of stone carving, the art of watchmaking, etc. And *artists* still claim for that reason a higher status than common handcraftsmen.

In recent times, England is likely to be the only country where handcrafts and arts are granted great honor. For royal princes do not shame themselves there to assume guild status. Even the Lord Mayor is chosen from the handcraftsmen, and for handcraftsmen to become members of parliament is very common.

§32 For most handcrafts in Germany, a youth must be an apprentice for three or four years before he is discharged (§ 8). In England, however, the

52 44) Compare with J. G. Heimeccius, *de Collegiis et Corporibus Opisicum, von Handwerkern und Zünften* [On handcraftsmen and guilds]. Halle 1723. J. G. Braumüller, *Über Gilden, Innungen und Zünfte* [On various kinds of guilds]. Berlin 1794. 8.

apprenticeship usually lasts seven years, thanks to an ordinance of *Henry the Eighth*, which was removed by *Queen Mary* and restored in the reign of *Elizabeth*. Almost always, young people enter the apprenticeship at age fourteen and are only let go in their twenty-first year. The apprenticeship years are relatively expensive, but in exchange the apprentices also enjoy liberties which are unthinkable in other lands.

The English handcraftsmen and merchants have united in corporations and trade societies which have their individual privileges and rights. They do not merely attempt to make these privileges and rights valid in various public circumstances, they even make an effort to increase them as much as possible. *London* has 72 such trade societies. Each one has its own headmaster, guild master, etc. These leaders are chosen annually from among the guild. They have to concern themselves with everything that pertains to the well-being and governance of the institute. At the head of all corporations there are the following twelve main trade societies: the silk merchants, the wholesale merchants or spice merchants, the cloth merchants, the seafood merchants, the gold- and silversmiths, the tanners, the tailors, the small wares merchants (*Kramwarenhändler*), the garment merchants (*Kleiderseller*), the iron merchants, the weavers, and the innkeepers. From these twelve trade societies the Lord Mayor is chosen annually, somewhat like a king of London. Since it is already known what power and degree of respect this magistrate possesses, I do not have to say much about it.

When kings and other rulers enter one of the aforementioned twelve trade societies now and then, then it is announced with much fanfare. If the *city of London* honors a well-regarded foreigner with the gift of its citizenship, then this foreigner is, upon receiving citizenship, required to enter one of those twelve societies. The prince of *Braunschweig* was rewarded for the important services he accomplished for the crown of England in the Seven Year's War with citizenship. The celebration organized for the event proceeded with great and well-earned pomp. The city of *London* gave this famous field commander, among other things, a large feast, and on that occasion he was presented by the Lord Mayor with the letter-patent for citizenship in a golden case. Out of politeness, the prince chose the guild of the spice merchants, since the Lord Mayor belonged to this group. At a later point, the king of *Denmark* also received citizenship and the privilege of an inhabitant of London. He then allowed himself to be taken up as a guildmember of the goldsmiths. The great statesman Pitt was in the guild of the spice merchants.[53]

53 45) "Bemerkungen über die Lebensart der Einwohner von London, den Handel und die Kaufleute" [Comments on the lifestyle of the residents of London, on trade and

The great *Catharina* of Russia also turned her attention to the handcraftsmen's groups in foreign lands and, according to their example, introduced several trade societies into her great realms. There are certainly still large lands where no guilds can be found, but in each case either the art itself died out, trading was predominant, or it was not possible to introduce any ordinance.

§33 The guild system, which certainly contributed so much to the improvement of handcrafts and fabricks, began at its very inception to degenerate in serious *abuses*. Most of the ceremonies, which took place when apprentices were discharged, on the occasion of becoming master, or at the election of the guild master, could even be seen as abuses. Thus many ceremonies were comprised, for example, in toasting, in roughhousing, or in hitting, etc., whereby some were injured, and some became coarse and debauched. The civil government soon took note and helped where it could. Regardless, many abuses still remain for the current age to remove.

Already in the year 1368, emperor Karl the Fourth charged the council and lay judges (*Schöffen*) at *Frankfurt am Main* with the task of doing away with abuses by handcraftsmen there and to dictate to them a series of ordinances for the common good.[54] From the electoral prince *Joachim von Brandenburg* there appeared in the year 1541, following an imperial police ordinance, another act. This act ordered the removal of handcraft abuses, which were comprised of taunts and slander, whereby one declared master and guildsmen or an entire guild dishonorable, as well as of forced drunkenness of the guildsmen and such things.[55] Further, many Swedish hatmaker journeymen tended to travel to Germany, and they were known as industrious and good people. In Germany they always enjoyed preferential treatment over the German journeymen. This annoyed the Germans. For that reason they crafted various plots against the Swedes. For example, as soon as a Swede arrived in Stralsund, he would be received with punches, and he would be told that he could expect such blows in all German towns. The Swedes were naturally driven away by this, and from that time onward, no Swedish hatmaker journeyman came to Germany. One thus had to do without various useful people who would have perhaps brought new methods and advantages with them.

---

merchants]. In *Journal für Fabrik, Manufaktur*, etc. [Journal for fabricks, manufacture, etc.] Vol. v. Leipzig. 1793. 8. P. 278.

54 46) Lünig's *Reichsarchiv* [Imperial archive]. P. sprec. Cont. Iv. Part 1. P. 588.

55 47) *Corpus Constitutionum Marchicarum* [Collection of the constitutions of the Mark]. Part v, section 2, chapter 10.

Such troublesome scenes became more and more common among German handcraftsmen. Many were also, as mentioned, occasioned by handcraftsmen's ceremonies. The planning,[56] polishing, preaching, baptizing, etc., the unnecessary handcraftsmen's greetings and ways of speaking, the severe expenses for being taken in by the guild, apprenticeship, discharging from apprenticeship, and achieving mastery: all these and other miserable habits were no longer to be tolerated. Therefore, they finally managed to bring about the famous *imperial law* from the year 1731, which was renewed in the year 1771, wherein they, with regard to the insolence of the handcraftsmen, put them back into their proper constraints and recommended to them order and good behavior.[57] This law has unfortunately only been maintained in a very few points, although it has been made publicly known in the German lands and territories and compliance with it has been ordered by edict of the territorial lords. Already prior to this imperial decree, the abuse of handcrafts had often been a topic of discussion at the imperial diet (*Reichstag*). The electorate of Hannover (*Chur-Hannover*) allowed a beneficial ordinance to be proclaimed

56 The Grimms' dictionary lists as one of the meanings of *hobeln* to punish by beating.

57 48) J. J. Schmaußen, *corpus iuris publici* [collection of public laws]. Leipzig 1794. 8. *Reichsschluß wegen der Handwerksmißbräuche* [Imperial decree on handcraft abuses] 1731. Ibid. P. 1680.

Imperial testimonies from Regensburg, on July 15, 1771, concerning the removal of handcraft abuses, particularly as concerns the so-called blue Monday. Imperial decree against the abuses of handcrafts from August 16, 1731. In the weekly Eisenach reports from 1752, no. 37f. Imperial testimony and imperial ratification of the commission degree concerning the removal of handcraft abuses. See *Oekonomische Fama* [Economic report]. Piece. 7. Frankfurt und Leipzig 1732. 8. S. If. *Kayserliche Verordnung die Abstellung der im heil. Röm. Reiche bey den Handwerkszünften eingeschlichenen Mißbräuche betreffend vom Jahr 1731.* Imperial ordinance for the removal of abuses in handcraft guilds in the Holy Roman Empire concerning the year 1731. Hamburg 1794. "Gedanken über den Reichsschluß de Anno 1731 wider die Mißbräuche der Handwerker" [Thoughts about the imperial decree of 1731 against the abuses of handcraftsmen]. In the *Eisenachschen vermischten Nachrichten* [Miscellaneous reports from Eisenach] 1753. No. 31. 32. C. G. Knorren, *Rechtliche Erläuterung des Reichsschlusses wegen der Handwerksmißbräuche* [Lawful explanation of the imperial decree against the abuses in the handcrafts] 1744. *Gutachten, die Abstellung der Mißbräuche bey Handwerkern betreffend* [Testimonies, concerning the removal of abuses in the handcrafts]. Jena 1746. 4. "Gnädiges Mandat zur Publicirung und genauer Beobachtung des Kayserl. Patents vom 23. April 1772 wegen Abstellung der Handwerksmißbräuche" [Merciful mandate on the publicizing and precise observation of the imperial patent of April 23, 1772, on the removal of abuses in the handcrafts]. Dresden September 18, 1772. In the *Leipziger Intelligenz-Blatt* [Leipzig paper] 1773. P. 57.

about this in the year 1692.[58] The same also occurred in the Brandenburg lands,[59] in Saxony,[60] and in the Austrian territories,[61] etc. To be precise, this matter became a topic at the imperial diet of Augsburg already from the year 1531 onward.[62]

Some harmful ceremonial activities still remain in various handcrafts: for example, ridiculing (*Hänseln*) comprised of pushing and shoving. The joiners in particular still practice it in various places and call it *hobeln* (from the *Hobel*). So too are *challenges* and *insults* fashionable in several places. People keep trying to root out this evil.[63]

---

58 49) *Churfürstl. Braunschw. Lüneb. Landes-Constitutionen und Verordnungen* [State consitutions and ordinances of the principalities of Braunschweig, and Lüneberg]. Hannover 1711. No. 93.

59 50) *Corpus Constitut. Magdeb.* [Magdeburg constitution] Capital 26, part II. P. 182.

60 51) *Corpus August.* Part. I. *Polizeyverordnungen gegen die Handwerksmißbräuche von 1543–1661* [Police ordinances against handcraft abuses from 1543–1661].

61 52) Franc. Autoni, Codex Austrisc. *Verordnungen wegen der Handwerksmißbräuche von 1567 bis 1689* [Ordinances against handcraft abuses from 1567 to 1689].

62 53) *Die ökonomische Fama* [The economic opinion]. Piece 6. Frankfurt und Leipzig 1731. 8. Page 3. "Von der Handwerker und Handwerkszünften in Teutschland Ursprung, wie auch Verfall derselben in allerhand Mißbräuche" [On the origin of handcraftsmen and handcraft guilds in Germany, as well as their decline in various kinds of abuse]. Page. 24. "Was zu Abschaffung der Handwerksmißbräuche von denen Kaysern und auf Reichstägen verordnet worden" [What has been decreed by the emperor and in parliaments for the removal of handcraft abuses]. P. 42. "Proiect was in den künftigen Reichsabschied wegen denen bey denen Handwerken eingerissenen Mißbräuchen zu bringen seyn möchte" [Project that could be brought into the future parliamentary recess against the abuses that have penetrated the handcrafts]. P. 66. "Was zu Abstellung der Handwerksmißbräuche in verschiedenen Reichslanden verordnet worden" [What has been decreed for the removal of handcraft abuses in various imperial territories]. Compare with *Europäische Staats-Canzeley* [European state chancellery]. Part. I. P. 126. J. G. Kast. *Specimina statutorum et consuetudinum, quae inter mechanicos vigent, irrationalium* [Specimens of statues and irrational customs which thrive among handcraftsmen] Strasburg. 1715. "Von den Handwerken" [On handcrafts]. In the *Hannöverischen nützlichen Sammlungen* [Useful collections of Hannover]. 1757. Piece 84.

63 54) F. Bock, "Über Mißbräuche und Unordnungen, so unter den Handwerksgesellen, besonders unter den Hutmachergesellen in den Preußischen Staaten noch üblich sind" [On abuses and disturbances which are still common in the Prussian states among handcraft journeymen, particularly among the makers of hats]. In the *Annalen der ökonomischen Gesellschaft zu Potsdam* [Annals of the economical society of Potsdam]. Vol. II. Issue 3. P. 18. "Benennung einiger Mißbräuche bey den Maurergesellen: 1) das Herausfordern 2) das Hundsfottschimpfen, 3) die Pfuscherey" [Description of a few abuses among the mason journeymen: 1) calling out, 2) cursing, 3) underhand dealings]. In the *Annalen der ökonomischen Gesellschaft zu Potsdam* [Annals of the economic society of Potsdam]. Vol. III. Issue 1. P. 146. Compare with C. H. Hiller *de abusibus, qui in Germania nostra in collegiis vigent opisicum deque medelis contra eosdem adhibendis* [On the abuses which are

§34 The poor treatment of the apprentices by some German handcraftsmen has also often been a subject of conversation, even in the imperial law mentioned in § 33, but unfortunately it has never been completely uprooted.[64] The reason for this is the prejudices of the masters, which were planted in them during their own formation, and which they now seek to impose upon the apprentice. In future times, this too will improve. For many states now begin to concern themselves with a better education and formation for those youth assigned to be apprentices, and also a better culture for the handworker's class (§ 35, § 42). The future generation can surely expect good things in this regard.

In England, no distinction is made with regard to rank or respect between apprentices and journeymen. In Germany, however, the step from apprentice to journeymen is held to be a very important one, and the journeyman has almost as much power over the apprentice as the master does. Journeymen, as assistants to the master, receive a certain compensation either weekly, monthly, or yearly. Since they are, as it were, hired by the masters, so were almost all

occurring in the handcraft schools of our Germany] Tübingen 1729. J. C. Dishmar, de collegiis opisicum apud Germanos emendandis [On improvements to the handcraftsmen's schools among the Germans]. Frankfurt 1731. *Nachricht von den Handwerksmißbräuchen* [Report on abuses in the handcrafts]. Frankfurt und Leipzig 1731. J. C. König, "Von Handwerksmißbräuchen" [On abuses in the handcrafts], in J. C. Siebenkees, *Beyträgen zum teutschen Recht* [Contributions to German law]. Part I. No. 3. P. 76. J. G. Sieber, *Abhandlung von den Schwierigkeiten in den Reichstädten das Reichsgesetz wegen der Mißbräuche bey den Zünften zu vollziehen* [Treatise on the difficulties in imperial cities to apply imperial law to abuses in the guilds]. Goslar und Leipzig 1881. 8. Additions to that in Schlözer's correspondence. Part. VI. P. 217f. J. U. von Cramer, "Von dem in Reichsgesetzen verbotenen Schelten, Schmähen, Auf- und Umtreiben bey Zünften und dem anmaßlichen Strafrechte oder Zunftzwange derselben" [On the scolding, reviling, heckling and disturbance in guilds, forbidden by imperial laws, and the proper criminal law or guild punishments of it]. In the *Nebenstunden* [Idle hours]. Part XXX. P. I. "Mandat der Stadt Hannover, das Schwärmen und Lärmen der Handwerksgesellen und Lehrjungen betreffend, vom 25. Oktober 1790" [Mandate of Hannover, with regard to the congregating and rowdiness of the journeymen and apprentices in the handcrafts, from October 25, 1790]. In Gatterer's *Technologischem Magazin* (technological magazine). Vol. II. Piece. 4. P. 255. "Anstalten wider Handwerksmißbräuche" [Measures against abuses in handcrafts]. In the *Leipziger Sammlungen* (Leipzig collections). Part. IV. P. 591. J. H. Pfingsten, *Von Handwerksmißbräuchen und ihrer Abstellung* [On abuses in the handcrafts and their removal]. Erfurt 1785. 8. – "Beytrag zur Geschichte der Handwerksmißbräuche" [Contribution to the history of abuses in the handcrafts]. In J. Beckmann, *Beyträgen zur Oekonomie, Technologie* u. [Contributions to economy, technology, etc.] Part. IX. No 3.

64 55) *Handwerksbarbarey, oder Geschichte meiner Lehrjahre. Ein Beytrag zur Erziehungsmethode teutscher Handwerker* [Handcraft's barbarism, or history of my apprenticeship years. A contribution to the method of educating German handcraftsmen]. Halle and Leipzig 1790. 8.

handcraft journeymen known, a few hundred years ago, by the name servant (*Knechte*),[65] as tailor's servant, carpenter's servant, metalworker's servant, etc. This has remained the case with a few handcrafts even now, for example, millers, bakers, brewers, cobblers, etc.

I will not judge whether this emergence of the name servant (*Knecht*) with handcraft journeymen is correct, or whether one can accept the following derivation as truth. In the fifteenth and sixteenth centuries, unmarried young people, even princes, were called *Knechte*. *Luther* even gave this name to kings, apostles, and prophets. Nobles were often called *Knechte*. Also the English word knight comes from this.

§35 Already since the origin of the guilds, the journeymen divided themselves into certain hierarchies, namely, into *young journeymen* (*Junggesellen*),[66] who have just finished their training, into *journeymen* as such, and into *foremen* (*Altgesellen*), who are closest to achieving mastery. Each of them had a small duty, for example, the foremen (*Altgesellen*) were directors of meetings, the young journeymen (*Junggesellen*) took over the invitations to celebrations, the transportation of bodies, etc. And thus it has remained for the large part until the present day.

Brotherhoods or groups of journeymen, brotherhood seals, journeymen courts, and such things have also been forbidden by the imperial edict of 1731. The removal of the *blue Monday*[67] and the self-determined removal from work in general was included as a particular subject of the imperial edict of 1772. However, one did not strictly obey this law everywhere and in all guilds. This is also the case with the *correspondence* of the handcraft journeymen, which the town authority should observe closely. The authority, however, stood in many cases free to make exceptions and particular dispositions.

The most important years of the journeymen are indisputably their *journey years* (*Wanderjahre*). The traveling or journeying of the journeymen is namely made law among the German handcraftsmen, so that they can broaden their knowledge and not immediately proceed to mastery. This purpose of journeying is certainly to be praised. But how many handcraftsmen make no abuse of their travels? How many journeymen do not wander around without plan and without usefulness in the world? How many become lazy and debauched

65 *Knecht* could mean slave, servant or bondman.

66 A word that later came to refer to a bachelor (*Junggeselle*).

67 "They say popularly, *blauen Montag machen*, to leave Work on Monday and go drinking and dancing (as Journeymen do), to play the Cobbler's Monday." Ebers, vol. 2 (1798), s.v. "Montag."

during the wandering? In the most recent times, one has taken the effort to seek out the advantages and disadvantages of wandering and the means whereby these disadvantages could be removed. Several learned societies, among others the royal society of sciences in *Göttingen*, in the year 1797, have made this topic into a prize competition, which was then executed satisfactorily enough. The main result from the consideration of several scholars ended up being that the wandering of the youths pursuing handcrafts for the enrichment of their knowledge in foreign places and to gain a better manner of life may certainly be useful, but that it must be directed by the state better than it had been done in the past and that, for example, the journeymen need to know which places are the best for the enrichment of knowledge in *their* handcraft, and not walk around for a long time useless in the world.[68]

In truth, several writings have been published for the good of the young craftsmen, which would be of the greatest use to them during their travels and for their formation in general.[69] But unfortunately, most of these people are

68 56) C. F. Mohin and J. A. Ortloff. *Beantwortung der von der Königlichen Societät der Wissenschaft zu Göttingen aufgestellten Preisfrage: Wie können die Vortheile, welche durch das Wandern der Handwerksgesellen möglich sind, befördert und die dabey vorkommmenden Nächtheile verhütet werden?* [Response to the prize question posed by the royal society of science at Göttingen: How can the advantages, which are possible due to the wandering of handcraft journeymen be encouraged and the disadvantages discouraged]. Erlangen 1799. 8. *Vom Wandern der Handwerksgesellen; eine Abhandlung aus der Gewerbpolizey und dem Handwerksrechte* [On the wandering of handcraft journeymen; a treatise from the industrial policing and handcraftsmen's law]. Nuremberg 1800. 8. C. L. F. Steffeck. "Versuch eines Beweises, daß das Wandern der Handwerksburschen nöthig und nützlich sey, daß es aber besser wie bisher vom Staate dirigirt werden müsse" [Attempt at a proof that the wandering of the handcraft journeymen is necessary and useful, that it would however be better directed by the state]. In the *Annalen der ökonomischen Gesellschaft zu Potsdam* [Annals of the economical society at Potsdam]. Vol. II. Issue 1. P. 108. "Wie können die Vortheile, welche durch das Wandern der Handwerksgesellen möglich sind, befördert, und die dabey vorkommenden Fehler verhütet werden?" [How can the advantages which are possible by the wandering of the handcraft journeymen be encouraged, and the mistakes which occur thereby by prevented?] by J. L. G. Leopold; im *Journal für Fabrik, Manufaktur* [Journal for fabrick, manufacture]. Vol. XVI. Leipzig 1799. April. P. 265 f. Also compare F. J. A. Kinderling, "Über die Wanderungen der Handwerksburschen und die damit verknüpften Gefahren" [On the wanderings of the handcraft apprentices and the dangers associated therewith]. In *Deutschen Museo* [German museum]. 1789. Vol. II. P. 123f.

69 57) For example, *Ueber die zu verbessernde Erziehung der Künstler und Handwerker* [On the needing to be improved education of artisans and handcraftsmen]. Nuremberg 1788. 8. Also in the *Journal von und für Deutschland* [Journal of and for German]. 1788. Piece 1. "Ueber den Ausstand der Handwerksgesellen zu Hamburg im August 1781, nebst einigen Reflexionen über Zunftgeist und Zunfterziehung" [On the walkout of the handcraft journeymen of Hamburg in August of 1781, along with a few reflections on the spirit of the

too coarse and too initiated in abuses and also already too possessed of prejudices to want to read useful books and teach themselves better. One can still hope for this with time, however, because many purposefully established civil schools, and in fact even many learned schools now concern themselves with a better formation for those youths determined to be handcraftsmen.

§36 For many years now in some of the trade societies, *gifts* have been introduced for the traveling journeymen. And from this the distinction between *gifted* and *ungifted* handworks has arisen. For some poor journeymen, this is certainly a nice relief. Others, however, make misuse of it. For this reason, many such gifts have been done away with, or are at most limited to 4 or 5 ggr[70] (15–20 Xr).[71] Through the so-called *surveying* (when journeymen inquire among various masters) one seeks to connect the journeymen arriving from afar to a place of work, so that they do not need to wander around themselves in a foreign place.[72] In England, the journeyman who wishes to have work allows his name to be written down in the public hall. There he is retrieved by the master who needs a journeyman. The most talented and industrious one receives the greatest compensation. He is not fed by the master, which is also the case in Holland and with most of the German handcraftsmen.

guilds and guild education]; in the *Journal von und für Deutschland.* 1791. Piece 7. P. 531f. *Der rechtschaffene Handwerker. Ein Buch für Handwerksbursche* [The proper handcraftsman. A book for youth studying handcrafts]. Second edition. 1796. 12. Almost the same thing: *Der rechtschaffene Professionist, ein Taschenbuch für alle Handwerker* [The proper professional worker, a pocket-book for all handcraftsmen]. 2 parts. Meissen. 1803. 12. (wherein many useful things appear). C. L. L. Steffeck, "Über Verfeinerung und Kultur des Geschmacks und der Arbeiten der Gewerkbetriebenden Staatsglieder" [On the refinement and culture of taste and works of the members of the state who are advancing industry]; in the *Annalen der ökonmischen Gesellschaft zu Potsdam* [Annals of the economical society at Potsdam]. Vol. III. Issue 3. P. 51. *Handbuch für alle Handwerksmeister, Gesellen und Lehrbursche* [Handbook for all masters of handcrafts, journeymen, and apprentices]. Potsdam 1784. 8. – *Wilhelm in seinen Lehrjahren und auf der Wanderschaft, oder Lehren, womit ein Lehrjunge seinen Verstand und sein Herz gut bilden, ein geschickter Meister und nützlicher Bürger des Staats werden kann* [Wilhelm in his apprenticeship years and on his journey, or lessons, whereby an apprentice who forms his understanding and heart well can become a talented master and useful citizen of the state]. Kempten und Leipzig 1792. 8.

70 Ggr = Gutergroschen, which was 1/24 of a Reichstaler.

71 Xr = Kreuzer.

72 58) "Abhandlung von Handwerksgesellen und geschenkten Handwerkern" [Treatise on journeymen and monetarily gifted handcraftsmen]. In the *Leipziger Sammlungen* [Leipzig collections]. Part V, p. 243. "Abhandlung von Handwerksgesellen und ihren Zünften" [Treatise on journeymen and their guilds]; in den *Leipziger Sammlungen.* Part VIII, p. 170, 183, 324.

When the three traveling years (often in the literal sense of the word *Wanderjahre*, through merely wandering around) have passed, then the journeyman can return to his place of origin and try to achieve the *right to mastery*. Some lazy journeymen have traveled in the world without having worked, have deliberately avoided all work, and have even resorted to begging. Useful measures taken against such useless people are greatly desired. In a few places, for example, it is in truth also forbidden to give those journeymen who refuse work offered to them the smallest gift.[73] The example of wise sovereign ordinances against guild abuses, such as the one from Fulda from August 31, 1784, contains something appropriate to this matter which deserves to be imitated more widely: "by the calculation of the journeyman's year, only that time should be calculated in which the wandering journeyman, according to the testimony of his written certificate, really engaged in work."[74] If only this ordinance were to be introduced in all other lands, much would be gained through that alone. The bad kind of traveling handcraftsmen would then take more care to find a respectable accommodation, they would come back more skillful, the horde of begging handworker youth, or rather tramps, would not plague the people in town and in the countryside so much, as unfortunately occurs now, even to the point of making houses unsafe.[75]

Certificates and the *handcraftsman's greeting* were introduced long ago in order to distinguish traveling journeymen from frauds. The *handcraftsman's greeting*, a saying that the traveling journeyman must ask when seeking a gift or employment, has been done away with by most handcraftsmen. The certificate, however, has been preserved and must continue to be. It is a testimony from that master with whom the journeyman has worked up until that point. It is confirmed by the magistrate and signed and sealed by the guild master and alderman (*Altmeister*) of the guild to which the handcraftsman belongs. Let one devote great precision and conscientiousness to this attestation and try to keep it from any kind of falsifications! In France, a small, printed book (*livret*) was introduced a few years ago as certification which contains all the laws and rules of behavior for the handcrafts' journeymen, and which has to

73 59) J. H. Fricke. *Grundsätze des Rechts der Handwerker* [Basic principles of law for handcraftsmen]. Göttingen und Kiel. 1778. 8. Paragraph 72.

74 60) *Journal von und für Deutschland* [Journal of and for Germany]. 1787. Vol. 1, p. 351, article 7.

75 61) "Vom Fechten der Handwerksburschen" [On the hazing of young handcraftsmen]. In the *Eisenachschen gelehrten Nachrichten* [Eisenach scholarly reports]. 1752. No. 7.

be shown to the civil governance before being signed and stamped.[76] There is thus no question of falsifications, and this practice certainly deserves imitation in other lands.

§37 The journeyman who has completed his traveling years and wishes to become master upon his return first has to prove his skillfulness through a *masterpiece*. Only in such places where the handcraft is *locked* – that means, limited to a certain number of masters – can he not achieve mastery at a time of his choice, even if he possesses all the necessary qualities for it. He then has to wait until a place opens up for him. The number of the *unlocked* handcrafts becomes at the present time ever smaller.

In the eighteenth century the question has occasionally been discussed of whether the *locked* or the *unlocked* guilds, of which the former are becoming increasing more frequent in Germany, are preferable.[77] To give a *decisive* answer to this question is difficult. With the locked guilds, one tries to limit the masters to a certain number so that the earnings are not spread among too many and are thus too greatly lessened for a particular individual.[78] The masters who are already guilded certainly have an advantage in the locked guilds, but the poor journeymen – who have survived their travel years, gained much knowledge, and would gladly now work on their own account – do not fare well. Also, the state and the general good often suffer through locked guilds. I could mention many examples where the most talented young people, who could have benefitted the state with their excellent new knowledge, went to a foreign land after being turned away and they pursued their craft with the greatest distinction. Only after many years, calculated from the establishment of locked guilds, where the young people see the difficulty of becoming master with their own eyes, can a more general advantage from the locked guilds be expected.

76 *Livret prefecture de Police. Extrait de la loi, relative aux Manufactures, Fabriques et Ateliers* [Prefecture booklet on police. Extract of the law relative to manufactures, fabricks, and workshops]. On the 22nd day of Germinal, year XI of the French Republic. 8.

77 J. U. von Cramer, "Ob eine Zunft auf eine gewisse Anzahl Meister eingeschränkt, und von einer solchen Verordnung appellirt werden könne" [Whether a guild can be limited to a certain number of masters] and In the *Metzlärschen Nebenstunden* [Metzler idle hours]. Part LVIII. P. 119. – J. G. Estor, "Vom Recht der daraus entspringenden Befugniß, die Handwerksmeister nach Befinden auf eine gewisse und geschlossene Anzahl zu setzen" [On the right of the authorization derived from opinion that the handcrafts' masters should be limited to a certain, closed number]. In J. G. Estor, *Kleinen Schriften* [Small writings] Piece 3. P. 388.

78 64) My *Handbuch der Technologie* [Handbook of technology], section 1 (Frankfurt am Main 1806). 8. P. 16 f.

§38 In the most recent times, when one wanted to decide in favor or against the locked guilds, one also began to argue more about the use and harm of the *guilds* (§ 8) themselves. Those who wanted to have the guilds completely done away with have naturally drawn the short straw. It is true that now in Germany no guild can be established without the approval of the police of the country (*Landespolizey*), and that on the contrary this institution is entitled to dissolve one or the other guild, or even all guilds in a land. But this causes many difficulties because of the imperial privileges granted to the guilds. In 1669, in the imperial diet, the suggestion to do away with all guilds in Germany was made, accepted by the majority of the imperial estates, but not put through. In the imperial decree of 1731, the handcraftsmen were at least threatened with this.[79] In England, where the guilds only have political purposes, no one has to become a member where he belongs because of his handcraft. A cobbler journeyman can, for example, become a guild member with the musicians, and then he can on his own account pursue his handwork as master (freeman).[80] No one can practice a trade, however, who does not belong to a guild. One attains mastery (or becomes freeman of the City) either through the seven accomplished apprenticeship years (§ 32) or through purchase. The purchase price varies according to the size of the city, but it seldom exceeds the sum of 30 pound sterling. No guild in England is locked. Every master has as many journeymen as he wishes, and masterpieces are unknown. Overall in the towns as well as the country side, every trade can be practiced. In France it is the same since the Revolution. The handcraftsman buys a patent for himself and is then allowed to practice his trade undisturbed. The greatest competition is certainly possible thereby. That through competition alone the handcrafts and arts are able to have climbed to the known degree of perfection, however, as the war councilor *Meyer* claims in his valuable treatise,[81] one need not admit. Herr Assessor of the Chambers (Karl Friedrich) *Wiesiger zu Treuenbriezen* has also given several important reasons to oppose the opinion of Herr Meyer.[82] France was already for a while the source of fashion, wherefrom the industry emerged in various forms. The genius of the French,

79 65) Reichsschluß (imperial decree) of 1731. Art. 6. – Moser, von der Landeshoheit in Polizeysachen. S. 199.

80 Appears in English in the original German, as does the phrase "freeman of the city" a few lines down.

81 66) *Annalen der Märkischen ökonomischen Gesellsch*aft *zu Potsdam*. [Annals of the economic society of the Mark at Potsdam]. Vol. III. Issue 2. Potsdam. 1802. 8. P. 60.

82 67) "Ueber die Nützlichkeit und Schädlichkeit der Zünfte" [On the usefulness and harmfulness of the guilds]; in den *Annalen der Märkischen ökonomischen Gesellschaft zu Potsdam* [Annals of the economic society of the Mark at Potsdam]. Vol. III, issue 2. P. 63 f.

especially for all things that catch the eye, guided them to many inventions which pleased due to their novelty, even if they did not bear the stamp of solidity with them as well. England gladly attracted foreign artists, was always possessive of its own good workers, had the fortunate situation for trade, possessed many excellent raw materials or could acquire them easily, rewarded useful inventions very honestly, and was in the position of delivering the main tools needed for most kinds of work, steel ones and iron, in the greatest quality. To that were added strict and conscientious *public repositories*, which one already had in Germany since the thirteenth century, although by now, in most regions of our fatherland, they require significant improvement. – Forced guild membership (*Zunftzwang*) can also no longer be seen as the main reason why in Germany some handcrafts attain a lesser degree of perfection than in England and France.

§39 Many regions in Germany in which forced guild membership (*Zunftzwang*) predominates can still claim that the handcrafts are in their finest bloom there. One can keep arguing as much as one will as to whether Germany should do away with or preserve the guilds,[83] but one should still vote for the latter, under the condition that the abuses occurring there (§ 33) should be removed as much as possible. We know that forced guild membership is a condition which excludes all those from pursuing a trade who have not learned it according to the laws of the guild and who have not sought to be taken in by the guild.[84] Naturally, many useful things must come of this. Future apprentices and journeymen will always be possessed by a greater ambition than unguilded youth. Unguilded journeymen will often proceed to mastery even if they are not diligent. Some capable youth will naturally be fortunate early on, if no obstacles are in his path on the way to mastery, while many others will be unfortunate early on. The immature master can expect the lot of all bunglers. Even if he can sustain himself for a while through inexpensive and poor work, people will soon get to know him, and then his trade will gradually dwindle away.

Too much competition certainly leads to less expensive prices, but usually also to less good wares. On the other side, one can certainly say: greater competition also demands a more noble rivalry. The *most talented* will then gradually

83 68) J. H. Firnhaber's *Historisch-politische Betrachtung der Innungen und deren zweckmäßige Einrichtung* [Historical-political analysis of the trade societies and their appropriate organization].

84 69) Compare with *Das wahre Beste der löblichen Zünfter und Handwerke* [The true best of the praiseworthy guilds and handcrafts]. By Dr. Reimarus. Hamburg 1770. 8.

receive the most work, and thus everyone will strive to perfect his work more, in order to outdo the others. But there are several unfortunate circumstances which turn some away from too much competition. Herr Meyer claims that more unrest and murders occur where guild brothers gather: where they drink, carouse, and all stand for one than with independent handcraftsmen, where everyone works for himself and goes on his own. Herr *Wiesiger*, however, counters him with the discipline that rules in the guilds, the authority of the guild master, the aldermen (*Altmeister*) and the foremen (*Altgesellen*), whereby already some riots were nipped in the bud. The drinking bouts, which have also been provoked by unguilded handcraftsmen, and where unrest often occurs, which as a matter for the entire citizenry, were more difficult to quiet.

§40 *Masterpieces* (§ 37) are certainly a very useful feature of the guilds. One has also found them to be extremely necessary in the establishment of the guilds so as not to accept every bungler into them. But in the ongoing culture of all handcraftsmen, and in the meaningful perfections that they undergo, the masterpieces have in the sense of a purposeful institution fallen greatly behind, and in recent times all sorts of abuses have found their way in which can by no means honor the guild. In almost all the handcrafts, a very old-fashioned, and often inappropriate piece is chosen for the masterpiece: one which the candidate for mastery has never seen, much less completed, a thing which no one can use afterwards. Small wonder, if such a masterpiece is not completed as desired! Several masters are present during the completion of the masterwork. These behave at times not completely honorably with the young man. They try to discomfit him with their conversation when he is already very nervous. Afterwards they find fault about such things on the completed masterpiece where no fault is to be found, and because of the least significant oversight they often destroy the entire masterpiece and thus rob quite a few talented young men of time, effort, money, and contentment. If only the state would watch carefully over the masterpieces and create standards in keeping with the current age and require the masters to remain conscientious and neutral in their observing and judging!

§41 Already several scholars of older times did not find it beneath themselves to go into the workshops of various handcraftsmen and be shown various things. They have even found it worth their trouble to think more about various kinds of work, techniques, and advantages. And in truth, they have in the meantime published results which could serve the perfection of the handcrafts and which people immediately used without further discussion. In the *eighteenth century*, many scholars have diligently applied themselves to the precise study of handcrafts, manufactures, and fabricks. A few have even made a particular course of study from them. People have in particular been sufficiently convinced of the use which these efforts have for handcraftsmen themselves, and which they must have for various men of business who occupy government offices. For many skillful techniques and the completion and use of many tools rely on *mechanical theories* which can be best learned scientifically. To the best choice, to the right use and to the most fitting, advantageous processing of materials belong, apart from accurate insights into *mechanics*, also broad knowledge of *the history of nature* (the study of animals, plants, and minerals), of *chemistry*, and of *physics*. One only recently came to appreciate that. In the workshops, all rules and customs were otherwise always passed down from master to journeymen and apprentices, and when these had achieved mastery, were then transmitted again to journeymen and apprentices, and they often rested upon faulty principles or none at all. Various scholars who gained knowledge of several handcrafts have, in recent times, through the help of their knowledge, indeed removed much that lay in the way of the greater perfection of these trades. As examples I need only to mention the tanneries, dye-works, and smelters. Certainly, there is still much to do here. But who is not well aware of the difficulties which oppose such an undertaking, and particularly the strong prejudices which one must struggle against!

The use for the general good which politicians gain from a proper look at the workshops of various handcraftsmen and artisans is, however, of very great importance. If they themselves see where it is lacking, then they can more certainly think of means of improvement, and thus help the flourishing of the state through handcrafts and arts.

§42 A single science has emerged from the knowledge of handcrafts, manufactures, and fabricks, which is taught at universities and other schools of higher learning. Herr Privy Councilor *Beckmann in Göttingen* first called it *Technology* in 1772, where one had otherwise, and incorrectly, said *History of the Arts* (*Kunstgeschichte*). In technology, not only the history of the handcrafts is taught. Rather, all works (*Arbeiten*), their results, and principles are explained thoroughly and clearly. It is also called *science of the arts* (*Kunstwissenschaft*) by a few of the most recent writers, because one obtains by it a scientific knowledge of all artisanal work (*Kunstgeschäft*). Technology is distinguished from art (*Kunst*) itself, just as knowing is from doing, or theory from praxis.

Once one began to pursue the knowledge of handcrafts scientifically and to educate scholars in a complete understanding of the various works in a number of workshops, then one came upon the idea of teaching craftsmen themselves technology and its auxiliary sciences. And from this there arose very useful *technological educational institutions*, or so-called *formational schools for handcraftsmen* (*Bildungsschulen für Handwerker*), *schools for handsraftsmen* (*Handwerksschulen*), and *Sunday schools*,[85] wherein were taught, in a comprehensible and universally intelligible (*gemeinverständlich*) way, drawing, and the basis of the study of nature and mathematics – principally *chemical* and *mechanical* principles – which are useful for so many handcraftsmen, alongside *chemical* and *mechanical techniques*.[86] In these schools, young handcraftsmen have to be very well prepared to understand various things much more easily, and even to discover various advantages, and to test and evaluate methods proposed by others. In various great cities like *Vienna, Berlin, Hamburg,* and *Munich,* such schools were erected in recent times which have already had the most fortuitous results. Let them be spread more and more in our German fatherland! How much would the state not gain from more thorough and skillful handcrafts! And how well would they not support the public, because the quality of the wares has to increase accordingly? In *Hamburg,* the free lecture for young handcraftsmen, artisans, and fabricants was created in 1791 by the praiseworthy *Society for the Promotion of the Arts and Useful Trades*. This society, which had its beginning in 1785, also offers annual prizes for the improvement of the handcrafts and arts, from which much good has also been accomplished 71).[87] And what has not already the *Society for the Encouragement of the Arts and Manufactures* in England achieved, just as the École Polytechnique has in France? The writings of the English *Society for the Encouragement of the Arts and Manufactures* testify extensively to the great progress which has been made in all handcrafts and arts in the eighteenth century.[88] In France, the *Society for the Encouragement of National Industry,* as well as the public exhibition of

---

85 Grimms' dictionary mentions that a *Sonntagsschule* (Sunday school) could refer either to one organized by the church for religious instruction, or one organized by the state to provide instruction in a particular trade.

86 70) Compare my suggestion for a state school for the arts, handcrafts, manufacture, agrarian economics and warfare. Tübingen 1777. 8.

87 71) J. A. Günther, "Über den Zweck, die Einrichtung und den Fortgang des von der Hamburg Gesellschaft zur Beförder der Künste und nützlichen Gewerbe veranlaßten unentgeldlichen Lehrvortrags für junge Handwerker, Künstler und Fabrikanten" [On the purpose, organization, and continuation of the free lecture for young handcraftsmen, artisans, and fabricants sponsored by the Hamburg society for the supporters of art and useful industry]; in the *Verhandlungen und Schriften der Hamburg. Gesellschaft zur Beförder der Künst und nützlichem Gewerbe* [Acts and writings of the Hamburg society for the supporters of art and useful industry]. Vol. II, p. 265. Also in the *Journal of and for Germany*. 1791. Part 2. P. 93 f.

88 72) *Transactions of the Society for the Encouragement of Arts and Manufactures*. London. Vol. I–XXIII. London 1783–1805. 8. – The *Repertory of Arts and Manufactures*. Vol. I–XXIII. London 1794–1805. 8.

many wares at *Paris* in the courtyard of the Louvre, created much enthusiasm for handcrafts and arts and contributed to their perfection more than a little.

Not only in recent times alone did one bring the thought of constructing a school for handcrafts into reality. *Nuremberg* had, already a few hundred years ago, a mathematical professorship where young handcraftsmen and artisans could receive necessary instruction in mathematics.[89] This instruction was naturally very incomplete compared to that of our age, but still it was useful enough. – *Nuremberg* supported workers who were sick or had returned due to an accident from the public treasury.

§43 Germany naturally still has much to do before its handcrafts, manufactures, and fabricks reach the level of perfection where these trades stand in England and France these days. But Germany, inventive Germany, which produced on its own so many new useful trades, and perhaps even the most useful ones that there are, can reach this level, can even climb higher, if it wishes to. There is no lack of proof for this fact. The most excellent handcraftsmen and artisans in England are the Germans. And should Germans and Germany not achieve this as well? Certainly, as long as their path to distinction is not blocked. Just as in those foreign lands, regents have to see that the education and encouragement of the handcraftsmen and artisans is maintained. They have to make sure that the most industrious and talented ones are awarded appropriately and that the inventor of useful things receives honorable distinction. In addition, they have to be certain that one cares for the just sale of German wares and attempts to weed out the harmful prejudice whereupon one claims that Germans can never attain the perfection of the foreigners in their work. Regents and government officials should occasionally inquire into the condition of handcrafts and arts, they reward where there is something to be rewarded, help, where help is required, do away with longstanding abuses in the guilds, and suppress all kinds of quackery, and through better inspections (§ 38) make effort to obviate all swindling, as far as possible. Then one can certainly expect a lot, and the continuous improvement of all these trades will certainly be the inevitable result. In a few German provinces, for example in Prussia and Saxony, one has already executed many of these rules, and seen the most beautiful fruits come to light.[90]

89 Petro Rami *Scholarum mathematicarum* [Branches of mathematics]. P. 65.

90 74) Compare with my *Vorschriften für Künstler und Handwerker, wie sie ihre Kunst mit leichter Mühe ausüben und die Produkte ihres Fleißes vollkommner machen sollen* [Regulations for artists and handcraftsmen, as to how they should practice their art with ease and make the products of their labor more complete]. Grätz. 1792. 8. – A. F. G. Voigt, *Beantwortung*

§44 One can always consider manufactures and fabricks to be the source of the population and wealth of a state, if their construction is based on reason and expertise. *Colbert* helped the French manufactures and fabricks flourish from the year 1665 onward. In Paris, he constructed the great mirror fabrick, a lace fabrick, and the tapestry fabrick (*Gobelins-Fabrik*),[91] and all those fabricks became even more famous in subsequent years. The removal of the Edict of Nantes in 1685, which had been in place since 1598, brought over a half million of the most talented and industrious people to foreign lands. Here they spread their arts with great eagerness. Here, they first introduced various manufactures or caused them to be accepted more readily. France had thereby given itself a grave wound, and Germany gained much. Thus one land was emptied of manufactures and fabricks, and the other was filled with them. Here trades collapsed, there new ones arose. Such an eternal teetering this way and that can certainly not be good. Not just single individuals, but also the state itself suffers from it. In England, all manufactures and fabricks were always built on the sturdiest supporting pillars. This also led to their relentless growth. At present, one estimates the number of people in this land who occupy themselves with the production of manufactured goods at two and a half million. *London* is the main repository of all English manufactures and fabricks. There one finds all possible kinds of cloths, fabrics, and materials, ribbons, lace, hose, hats, steel-ware, things made of ivory, wood and clay, watches, and the like. One can scarcely think of any ware that is not to be found in *London*.

Not every place or every province is suitable for every kind of manufacture or fabrick. In one place this one can flourish more, and in another place that one. This has long been known by smart government officials and smart manufacturists. On the other hand, the ignorant who did not know or want to know this often end up badly, and have not infrequently caused themselves the worst ruin due to the poor construction of manufactures or fabricks without a plan. I could give many examples of uninformed speculators, who wish to introduce a new trade in a place or a land, but their unreasoned, and often senseless plan collapses, as was easy to anticipate, and pulls the entrepreneur with it into the abyss. Wherever, in particular, manufactures or fabricks should be constructed, all primary and secondary materials must firstly be there in sufficient amount,

---

der Frage: durch welche Mittel könnten unsere Handwerker dazu gebracht werden, daß sie diejenigen Verbesserungen ihrer Gewerbe nützen, deren Zuverlässigkeit durch die Erfahrung der Ausländer erwiesen ist? [Answer to the question: through what means could our handcraftsmen be brought to use those improvements in their trade, whose reliability is proven by the experience of foreigners?]. Erlangen 1799. 8.

91 75) *Histoire de France sous le regne de Louis XIX* [History of France under the reign of Louis XIX]. Volume 3. Rotterdam 1718. 12. p. 360.

and economically, the wages should not be too high, and one cannot lack in sufficient sales and an easy method of shipment. Smart manufacturists or fabricants have known this for a long time. But how idiotically the person acts who tries to construct glassworks in a place poor in wood and coal, or who tries to construct a linen manufacture in a region where no flax grows? And yet, how many a one has not followed such a poorly constructed plan and has had to become wise through suffering losses! Also, smart people have always consulted the nature of water and air. What an important influence does not water, for example, have in bleaching, dyeing, papermaking, and soap boiling? And how harmful is not the effect of the air which is hot and dry on dyed fabrics, if these are prepared in southern climates?

§45 In earlier times, all fabricks belonged to the handcraft, and the merchant was merely someone who financed and promoted the handcraft. With cloth and fabric manufactures, one usually still follows this practice. In these cases there were nothing but industrious and talented workers whose trade remained in its greatest bloom for a long time. Gradually, however, in many places the merchant began to elevate himself to master and, in exchange for wages, to employ journeymen whom he, because of other business concerns, observed only infrequently, and whom he, because of a lack of expertise, could seldom correctly observe and evaluate. One therefore does not need to wonder why some work was produced so hastily and imperfectly. Those owners of manufactures and fabricks who lacked their own thorough expertise in this area always did well when they hired a really honest master handcraftsman, or at least a knowledgeable director who did not only keep the workers properly busy, but also kept an eye on the quality of the finished work, and who at times knew how to find new advantages.

The state which wishes to help manufactures and fabricks flourish as much as possible in a country also concerns itself with the proper sale of wares. It requires domestic wares to have only very few taxes and places a burden on the entrance of foreign wares through higher taxes. We can thank just such an institution for the fact that handcraftsmen, artisans, and fabricants practice their trade eagerly, and that they invented all kinds of advantages and numerous machines, whereby the work took place more easily, in less time, and yet more perfectly than before, where less oversight motivated it.

Who dares even to answer the question affirmatively: has every handcraft already the most appropriate means for its works? Certainly there is much more left to improve and to invent. All handcrafts and arts are certainly less improved by completely new inventions, but they are astonishingly improved through the translation of means from one workshop into another. Many handcrafts, as different as their materials and wares are, accomplish various works for the same purposes, and require for that the same techniques, the same instruments, and the same machines and other means. For example, how many kinds of work there are whose purpose is to make something smaller, or to flatten it, or make something smooth, or rough, or granulated, etc. Not all works have the same means for this, and yet some could have an advantage over others which one neither has nor is familiar with. The fine suggestion from Herr Privy Councilor Beckmann *for a general technology* or an inventory of all related works, instruments, and machines certainly deserves to be realized soon.

§46 Many handcraftsmen, artisans, and fabricants have elevated themselves and kept their trade flourishing in recent times by knowing how to conform to the whims of fashion, and by understanding how to keep tempo with fashion through a small adjustment in their works. The English have come particularly far in this art. How often does one not hear, for example, about a new English cotton cloth that is eagerly purchased, yet it basically does not distinguish itself much more from the old one except through name? Other craftsmen, artists, and fabricants, on the other hand, who for the most part keep working in the same old way, lag behind and ultimately fall completely to ruin. The public likes – and always has liked – things which are new and things which are good at catching the eye, or which possess never-before-seen artistic touches. The English, who are very inspired in the invention of such things, soon realized this to their advantage. One gave them patents for such inventions, or the right for their individual development over a span of years. That naturally had to have had a consequential influence on the national industry and national fortune. So many patented things have appeared in recent years: patented artful candle snuffs, patented shaving knives, with which one cannot cut oneself, patented scissors and the like, that the unpatented things fell almost completely out of view. Germans would have certainly made some of those inventions just as well or even better. But one gives them no patent for it, because one does not support them as in England. And in fact, already many German inventors have brought their inventions to England, where they are naturalized. We Germans now receive various German inventions as foreign imports, praise the inventive spirit of the proud islanders, and complain about German apathy.[92] In England, even the artisan who is not well-to-do can more

92 77) "Fragmentarischer Beytrag zur Beförderung des Fabrikwesens in Teutschland" [Fragmentary contribution for the furtherment of fabrick matters in Germany]. *Journal für Fabrik* [Journal for fabricks]. Vol. 11. Leipzig 1798. 8. October. p. 321.

easily apply his invention in fabrick work. For through his patent he is secure from all foreign encroachment, and the invention receives, as it were, the value of a real estate. The invention has to have no value at all if the inventor may not count on secure sale and profit. As keeper of a patent, it is easy for the inventor to receive a monetary advance from a capitalist. In this way, for example, the poor potter *Wedgwood* became so extraordinarily wealthy through the invention of his earthenware, and the poor barber *Arkwright* through the inventing of his spinning machines.

In Germany this is completely different. An invention by a German is immediately imitated as soon as one learns of the new ware. If Germany's regents cared at all for such regulations, then industry would rise to a much higher level in our fatherland as well. Germany would finally no longer be dependent upon the industriousness of the English, new fabricks would open up and flourish beautifully, and where now unemployment and poverty rule among the inhabitants, there would finally be a deficit of human hands, and this would most surely lead to the increase of available machines and the invention of new ones. Perhaps a future history writer of my science will be fortunate enough to praise such a period in our German fatherland.

§47 It is clear that the invention and application of useful machines contributed the most to the improvement of manufactures and fabricks (§ 24). England and France offer the most convincing proof of that. Whoever, for example, knows about the establishment and great extent of the fabricks system in England will never wish to dispute how beneficial machines are. For example, how far back would cotton manufacturing still be, if no machines had been invented for it!

In any case, England has machines alone to thank for the glorious flourishing of its manufactures and fabricks, and for the sale and real value of its wares. In England, all groceries are very expensive, and the English wares could certainly not compete with others if one had not prevented this misfortune with machines. England's fabricks and manufactures had fallen, the bounty (*Ausführprämie*)[93] notwithstanding, and without the the beneficial *machines* the inflow of national wealth and of national earning would have dried up. Certainly, a hundred thousand hands, which are now sought for and which find occupation, would be reaching out for work in vain.

Also, in several areas near the Rhein in Germany, the prosperity and population is completely a work of machines. These also give proof that they by no

93 "Bounty" here is a term in economics that referred to the bonus paid to encourage trade.

means take the place of the hands of men, but rather bring them alongside, and that they elevate the general wellbeing of the land in precisely the measure as they satisfy the self-interest of the speculating fabricant.

In the beginning, the introduction of machines was seen as detrimental and dangerous. In particular, one was concerned that unemployment and deep poverty would ensue in the lowest class. The builder of machines in England thus had to suffer much persecution at first. People's discontent soon calmed down, however. From day to day, more and more new and better sources of activity opened up. Human hands, which were in the beginning really suppressed, soon had to be summoned in greater number, because all wares made by machines turned out much more beautiful and economical. This increased their sale tremendously: it increased the number of machines, the number of fabricks, and subsequently too the number of necessary managers and workers, who still had to be kept active alongside the machines. Thousands of people now exchanged a hard fate for a milder circumstance, and thousands of needy people became more fortunate. People were pulled away from the rough and exhausting manual labors, which gave them a miserable income, and shown an easier and more rewarding occupation. The initial complaints turned into rejoicing.

One certainly has to accept that the invention of the plow, the scythe, all kinds of mills, and other machines provoked many similar complaints at first. But soon one admitted the beneficence of these inventions and all complaints were cut off at the root. That several German fabricks owe their prosperity exclusively to machines is as far beyond doubt as the fact that machines in no way suppress human hands, but rather attract them. The more economical price of wares that the machines managed to accomplish soon expanded the sale of wares so greatly that domestic and foreign consumption more than doubled, and the damages due to the unemployed workers were soon richly repaid. As examples, I will here mention the fabricks in *Elberfeld* and *Gemarck* (or in *Barmen*). These too have only the ingenious machines working in these places to thank for their state of prosperity. They are an eloquent proof of the irrationality of those friends of humanity who advised against the use of machines because they will deprive workers of their bread. Precisely because those fabricks are populated by machines, several thousand men find work and financial support. If one were to remove the machines from the fabricks *today*, then *tomorrow* thousands of workers will be without income. The products will disappear with the machines, and the region will soon be changed into an empty wasteland.

§48 Those who invented, applied, or facilitated the application of machines acted under no circumstances wrongly and unfeelingly towards their needy fellow men. For now, these men could often reach for another source of sustenance that made them more useful for the whole and made what had up until that point been a hard fate noticeably milder. In any case, those people deserved the warmest thanks from the state and all humanity who removed those labors from human hands which an unreasoning creature (*unvernünftiges Geschöpf*) or a lifeless force could accomplish at far less expense and in a shorter time. Every discovery or invention could not be praised enough for easing the coarse manual labor or the preliminary processing of raw material. These inventions, regardless of the purpose, saved many human hands and kept them free for more useful occupations and contributed to lowering the cost of the most urgent and, for the working population – in particular for the poor manual laborers – the most essential life-needs at a lower price, also by repressing foreign artisanal products and saving amounts of money which previously foreign countries had received through the sale of their wares.

Through the application of useful machines there also arose so many meaningful advantages for the state that it would be idiotic to doubt them. And one may brazenly claim that even the small fabrick revolutions that emerged through the introduction of machines laid the groundwork for many other manufactures and fabricks that one otherwise would not have thought of. Smart fabricants know that they have to reduce the price of their wares just as much as the rising value of groceries would drive it up if they otherwise want to remain flourishing. For now the workers naturally demanded a higher wage and for the encouragement of sales one cannot think of raising the price without running into the danger of seeing the rich affluence all at once directed to other fabricks. Otherwise, one had, for example in *Elberfeld* and in *Gemark*, machines which completed twelve pieces of ribbon at one time. Soon one expanded them to twenty-four. Now there are those which produce forty or more pieces at one time. And so too with all other kinds of work. There always have to be reasons, of course, before one moves to the actual use of machines. I have explained such reasons clearly enough in prior sections.[94]

---

94 78) Compare with my "Vorurtheile über Maschinen und Menschenhände" [Prejudices about machines and human hands]. *Journal für Fabrik* [Journal for fabricks]. Volume VIII. Leipzig 1795. 8. April. Page 262. "Gedanken über die Anwendung der Maschinen, wodurch Menschenhände erspart werden" [Thoughts about the application of machines, whereby human hands will be spared]. *Journal für Fabrik*. Volume 9. Leipzig. 1795.

If the fabrick industry enlivened all lands to an equal degree, the fabricks would only have to limit themselves to the needs of their own countries, and then no new increase in foreign trade would be expected. But this situation is unlikely to occur. For the characteristics of nations, the soil of the various kinds of countryside, the composition of their products, and the like, is much too different for one to expect the identical flourishing in various trades.

Many machines in manufactures and fabricks are constructed such that children, who otherwise run around uselessly, or women and old men could accomplish the work on them, which is often comprised in nothing other than in laying down materials, in standing by, and in a proper oversight. Already in recent times, primarily since the middle of the eighteenth century, one has begun, wherever there are unoccupied poor children and other weak people, to create *industry schools*, wherein those poor people together take on all kinds of easy but useful works. For example, they card wool, spin yarn, braid and weave blankets, make winter shoes and baskets, and also produce things out of paste. Among other places, *Göttingen* and *Frankfurt am Main* can boast of such institutions which make the best progress, free the state from having beggars hanging on its neck, and transform them into useful members of human society. The industry school of the superintendent *Wagemann* in *Göttingen* has produced beautiful fruits from the beginning, and through the present day it has been very beneficial to other German states.[95]

In many lands, one now allows various raw materials, vegetables, fossils, and the like, unknown through now, to be sought and thereby puts idle hands to work. Herr Superintendent *Schäffer zu Regensburg* was the first German who troubled himself to make the down ("pappus" or *Saamenwolle*) of a few native plants, various plant stems and tree bark suitable for the preparation of fabrics. He also discovered in many plants useful surrogates for the matter used for the preparation of paper. He himself made very many experiments which were successful. And if these materials do not, in fact, deliver as good fabrics or as good paper as the usual, partially foreign, products, one could still use idle hands for their refinement and perhaps prepare average quality wares for ordinary use from them. In any case, one should pay more attention to this matter than one has in most recent times. One should not repay German creativity and German industry so poorly that their fruits are cast so uselessly to the

P. 366f. "Fragmentarischer Beytrag zur Beförderung des Fabrikwesens in Teutschland" [Fragmentary contribution for the furtherment of fabrick matters in Germany]. *Journal für Fabrik*. Vol. 15. Leipzig 1798. October. P. 321f.

95 79) Arnold Wagemann, *Über die Bildung des Volks zur Industrie* [On the education of the populace for industry]. Part. 1. P. 228f.

side. One could certainly still achieve some advantage from them, even if it is not the greatest.

§49 Almost all manual works which demand mechanical or chemical skills come with a danger whose elimination or reduction would have the most beneficial results. Often severe external injuries lead immediately to death, as is the case with miners and metalworkers, millers, roofers, many kinds of fabrick workers, etc. Often, danger is induced by harmful material which one seeks to acquire or process, as by the excavation of vitriol, alum, sulphur, arsenic, and quicksilver, in gold-plating, leather tanning, coppersmithing, white lead painting (*Bleyweismahlen*), plaster milling (*Gypsmahlen*), etc. Naturally, one has long since taken various kinds of preventative measures against such dangers, but these measures were in part not entirely safe, and in part incorrect and poorly chosen. With most kinds of labor one did not even know the cause of many bad effects and thus kept working without any kind of safety measures (*Vorbeugungsmittel*). The Italian *Ramazzini*[96] was, as far as I know, the first who, prior to the middle of the eighteenth century, with no little success, called attention to the many kinds of work done by manual laborers and artisans which were often accompanied by dangerous illnesses and not infrequently a sudden death. He also, on the other hand, claimed various means of preservation and rescue which the well-deserving German doctor *Ackermann* justifiably opposed.[97] Subsequently, Sirs *Adelmann* and *May* have also called attention to the same subject and shared all kinds of medicinal suggestions whose observance is very much to be desired.[98] I have, in my prize essay, observed this matter of such importance for humanity from the mechanical and chemical perspective and also proposed means which are to some degree new, to prevent such dangers and to make the ones already present harmless, or at least less harmful.[99]

96 Bernardino Ramazzini (1633–1714) was an Italian physician.

97 80) Bernardino Ramazzini, *Abhandlung von den Krankheiten der Künstler und Handwerker* [Treatise on sickness of artisans and handcraftsmen]. Newly revised and expanded by J. C. G. Ackermann. Two volumes. Stendal, 1784. 8.

98 81) G. Adelmann, *Über die Krankheiten der Künstler und Handwerker* [On the sicknesses of artisans and handcraftsmen]. Würzburg 1803. 8. – F. May, *Die Kunst die Gesundheit der Handwerker gegen die Gefahren ihres Handwerks zu verwahren* [The art of protecting the health of handcraftsmen from the dangers of their handcraft]. Mannheim. 1803. 8.

99 82) J. H. M. Poppe, *Allgemeines Rettungsbuch, oder Anleitung vielerley Lebensgefahren, welchen die Menschen zu Lande und zu Wasser ausgesetzt sind, vorzubeugen, und sie aus den unausweichlichen zu retten* [General rescue book or guide for preventing various mortal dangers which people are exposed to on land and in the water, and for rescuing them from the unavoidable ones]. An award-winning prize essay. Hannover und Pyrmont 1805.

Still, care is taken in only a very few lands by the state to make sure that the dangers faced by workers is correctly portrayed, and that the means of prevention or rescue are really taken to heart. External injuries and thereby often a sudden death have, for example, often occurred in gunpowder mills due to burning, in other mills through the bursting of a millstone or in various chemical fabricks through violent explosions of vessels, etc. ... Internal injuries and the most frightening dangers are often produced by the harmful effects of the material being prepared, for example by breathing in harmful fumes or a harmful dust, whereby the air is infected, like arsenic vapors in the preparation of platinum, quicksilver vapor in the process of gold-plating, lead vapors in the glazing of dishware, the dye dust from the process of rubbing dyes, the ceruse dust, gypsum dust, lime dust, in the painting of ceruse, gypsum, and lime, the vapors of beer, wine, and from the tannery, etc. I will faithfully list the most effective mechanical and chemical means against this, according to both familiar and new ideas, in the history of the individual handcrafts.

§50 In recent times, one has tried in different ways to divide technology neatly into various compartments, and then to treat it scientifically. Herr *Jung* thought the easiest and most appropriate division was the one where the preparations, which were caused by the same or similar forces, were arranged from easier to more difficult ones.[100] He soon realized on his own account, from another perspective, how difficult this order was because such a great difference can occur in several processes. He therefore constructed another foundation to his system of technology, in that he divided all handcrafts into *mechanical* and *chemical*. The mechanical handcrafts he further divided into *preparations by hand* and *by machine*. To the first he relegated those which occur by hand using simple tools, to the latter category, however, the processes of milling and weaving. Chemical handcrafts he separated into *chemical preparations using moisture* and *chemical preparations that are dry*. In the first he placed the preparation of salt, fermentations, distillations, boiling processes and bleaching. To the latter, however, belong distilleries, smelteries, and metalworks. This division by Jung has been followed by other technological writers,

8. P. 223f. The first edition published by the prize organizer, the famous Herr Grafen von Berchtold, was published in 1804, in Vienna.

100 83) J. H. Jung, *Versuch eines Lehrbuchs der Fabrikwissenschaft* [Attempt at an instruction book on fabrick science]. Nuremberg. 1785. 8. Second edition. Nuremberg. 1794. 8. P. 6f.

such as Herr *von Lamprecht*,[101] Herr *Rössig*,[102] and Herr *Gotthard*,[103] with scant change.

Herr *Rosenthal* divided all the manual laborers into *ten classes*.[104] In the *first* he placed those who provide nourishment for the body; in the *second* those who clothe us, who deliver the material for it and prepare the articles of clothing; in the *third* those who take care of our dwellings; in the *fourth* those who make all kinds of household utensils, vessels, and the like; in the *fifth* such ones who deliver other necessities; in the *sixth* those who care for our health; in the *seventh* those who consider our comfort; in the *eighth* such people who concern themselves with our amusement; in the *ninth* those who protect us; and in the *tenth* all those who prepare the means for us to become more prudent and wise.

§51 Still, technology has been given no scientific partition that is precise and error-free. To Herr Privy Councilor *Beckmann* it seems the most advantageous to bring those handcrafts whose main works are identical or similar into a single section and certainly such that the simpler ones are presented first and the more complicated ones afterwards. In this manner he presents a list of 51 classes,[105] but deals with the handcrafts and arts according to the three kingdoms of nature, which provides him with three main divisions of technology.

Herr *Walther* also uses the three realms of nature as the basis for the main divisions of the handcrafts. He thus divides them *firstly* into such which refine products of the animal kingdom; *secondly* into such which process products of the plant kingdom; and *thirdly* into such products that process or refine products of the mineral kingdom. He continued to further divide the handcrafts of each of these divisions according to the various classes of natural products.[106] Thus he allowed, for example, in the classification of handcrafts of the first category, first those handcrafts to follow which take their materials for processing from the class of mammals, and then those which process materials from the

101 84) G. F. von Lamprecht, *Lehrbuch der Technologie* [Instruction book of technology]. Halle 1787. 8.

102 85) C. G. Rössig, *Lehrbuch der Technologie* [Instruction book of technology]. Jena 1790. 8.

103 86) J. C. Gotthard. *Handbuch der praktischen Technologie* [Handbook of practical technology]. Hamburg und Maynz. 1804. Part 1. P. 9f.

104 87) J. C. G. Jacobson. *Technologisches Wörterbuch* [Technological Dictionary]. Continued by G. E. Rosenthal. Part 5. Berlin und Stettin. 1793. 4. Preface.

105 J. Beckmann, *Anleitung zur Technologie* [Guide to technology]. Göttingen. 1777. 8. Sixth edition. 1806. 8.

106 89) F. L. Walther, *Versuch eines Systems der Kameralwissenschaften* [Attempt at a system of the cameralist sciences]. Part 3. Technology. Gießen 1796. 8.

class of birds, etc. He therefore came closer to a general division of technology. Only, too bad that one had to separate some handcrafts from another in this classification, which because of the great similarity in work could have much more properly followed one another closely, for example, wool weaving and silk weaving relative to cotton weaving and linen weaving.

Then, Herr *Bensen* appeared with his new division.[107] He believed that among all of the handcrafts, those should come first in order which make raw or already prepared natural products fit for further use in an *easy and simple manner*, and which demand less effort and fewer auxiliary means than the other handcrafts. He therefore calls them *handcrafts of the first order* or *handcrafts of simple processing*. These he divided further into three smaller sections: into *handcrafts in the narrower sense*; into *mechanical works*, which may be practiced with help from actual machines; and into *chemical works*. The second main class, which he refers to as *handcrafts of the second order* or *handworks of complicated processing*, should contain those which require *more artful* and *difficult* auxiliary means for their execution, just as they also demand *more intellectual effort*. But this kind of classification was still very unhelpful and lacking and stood far behind the earlier ones.

The most recent classification of handcrafts known to me stems from *Christian Rösling* and *Carl Wilhelm Rösling*.[108] They divided technology into *two main sections*. The *first section* deals with those *handcrafts, in which the essence of the artisanal product is not comprised of certain forms attached to it, but rather in certain chemical qualities of the material*. In twenty-four main sections, the following subjects appear: 1) the preparation of alcoholic drinks, 2) baking and drying, 3) the preparation of chocolate, 4) butter- and cheese-making, 5) the preparation of various acids, 6) the preparation of alkalis, 7) the preparation of various middle salts, 8) the making of tar pitch, pine oil, and lamp black (*Kienruß*), 9) oil pressing, 10) the preparation of soap, 11) the production of white starch and powder, 12) the preparation of various dyeing materials, 13) waxing, bleaching, dyeing and varnishing, 14) tanning, 15) the preparation of lime and isinglass, 16) the production of phosphor, 17) the preparation of sulphur, 18) the production of tinder (*Feuerschwamm*), 19) the burning of lime, gypsum, and coal, 20) metalworking or the processing of ore in metals, 21) steel burning, 22) cooking, 23) the apothecary art, 24) the

107 90) H. Bensen, *Versuch eines systematischen Grundrisse der reinen und angewandten Staatslehre* [Attempt at a systematic outline of pure and applied political theory]. Vol. 3. Erlangen 1799. 8. P. 108f.

108 91) Chr. Lebr. Rößing und J. C. W. Rösling, *Kompendiarisches Handbuch der Technologie* [Concise handbook of technology]. Vol. 1, section. 1. Erlangen 1806. P. 42f.

changes through which living bodies are given other qualities, like the carving and killing of animals. The *second section* is concerned with those *handcrafts, in which the essence of their artisanal products, apart from certain chemical qualities of their material, is comprised particularly of associated forms.* This section falls into *three chapters*. The *first chapter* takes up those handcrafts in which separation is the main act of production; the *second* those handcrafts, in which the joining of similar and dissimilar materials is the main act of production; the *third*, however those in which neither a separation nor a joining is the main act of production. The handcrafts which produce a separation through crushing, grinding, and rounding are relegated to the first paragraph of the first chapter. To the *second chapter*, however, belong those which have the purpose of dividing by pulverizing or smashing to pieces, and thus follow, one after the other, in the *third paragraph* separation by grating; in the *fourth* separation by carving into pieces; in the *fifth* the separation by splitting, slashing, tearing, hacking, hewing, carving out, cutting away, carving, clipping, cutting into, cutting out, pricking out, gouging out, etching with acid, boring out, skimming off, rubbing smooth, scraping, grinding and polishing; in the *sixth*, separation by woodturning. The handcrafts in the *second chapter*, which combine materials together, represent, like those in the *first* paragraph, joining by binding; in the *second* paragraph, however, a binding by twining; in the *third* paragraph binding by covering with yarn (*Ueberspinnen*); in the *fourth* paragraph binding through sewing and braiding; in the *fifth* paragraph binding through lace weaving (*Klöppeln*); in the *sixth* paragraph binding through tapestry weaving (*wirken*); in the *seventh* paragraph binding through (cloth) weaving; in the *eighth* paragraph binding through the application of ink, dye, and wax; in the *ninth* paragraph binding through gold-plating, silverplating, tinplating, application of foil and enamel. In the *third main section*, where handcrafts appear within which neither a separation or joining alone can be taken as the main act of production, the *first* paragraph has to do with the production of noodles in bulk; the *second* paragraph with the preparation of tobacco; the *third* paragraph with fulling (*walken*) and felt-making; the *fourth* paragraph with paper and paste-making; the *fifth* paragraph with the fabrication of various things from paper and plaster; the *sixth* from various manufactures of curried or woven cloths; the *seventh* from the most superior works in felt, hair, and feathers; the *eighth* from the fabrication of various things from tallow, wax, and dough; the *ninth* from the preparation of various artificial flowers; the *tenth* from the various mosaic works; the *eleventh* from the preparation of various products from calk, gypsum, and clay; the *twelfth* from glass-making; the *thirteenth* from various works in glass; the *fourteenth* from the preparation of black pencil or red pencil; the fifteenth from the preparation of gunpowder; the *sixteenth* from the preparation of various things made of metals (contains conceivably many handcrafts, namely all of the ones which from base (*unedlen*) metals manufacture the most superior cast-wares (*Gußwaaren* sic), from noble and base metals the welded and embossed (*getriebenen*) wares, the thin sheets of metal, the wire, thin plates of metal, and such things); the *seventeenth* from clock-makers, the *eighteenth* from the preparation of bellows; the *nineteenth* from the preparation of the most superior musical instruments; the *twentieth* from the preparation of fire pumps (*Feuerspritzen*); the *twenty-first* from the preparation of fans, parasols, and umbrellas; the *twenty-second* from Nuremberger toys; the *twenty-third* from the coopers; the *twenty-fourth* from the waggoners (Wagner); the *twenty-fifth* from the joiners (*Schreiner*), the *twenty-sixth* from the carpenters; the *twenty-seventh* from the masons (*Maurer*); the *twenty-eighth* from the plasterers; the *twenty-ninth* from the woolen cloth-makers (*Kleber*); the *thirtieth* from the roofers; the *thirty-first* from the tilers (*Schornsteinfeger*); the *thirty-second* from the preparation of surgical instruments; the *thirty-third* from the preparation of mathematical and physical instruments, and the *thirty-fourth* from the changes whereby certain forms are maintained on inspirited bodies (*beseelten Körpern*).

One sees here with this classification however many well-chosen connections, especially in the twentieth main division (*Hauptabtheilung*), but just as unmistakable are also the inadequacies of these orders. An entirely error-free, precisely-agreeing, complete and systematic division of technology will likely for a long time still (perhaps always) remain among the devout wishes.

§52 For a long time, people have wished for a complete work wherein all handcrafts and arts would be extensively described and depicted with all their basic principles, techniques, tools, and machines. That such a project is extraordinarily complex, extensive, and beyond the forces of a single man is easy to admit. A pair of very talented and erudite men, *von Reaumur* and *Shaw*, actually took steps to execute such a work. In particular, the former shared his plan with the French Academy of Sciences, and this group was the one which, supported by many experienced men, allowed the plan to be brought to execution. The extensive description of the arts and handcrafts, which was brought to light by the academy, belongs to the most important scholarly works of the eighteenth century, even if it does not comprise a completed whole.[109]

109 92) *Descriptions des Arts et Metiers, faites ou approuvées par Messieurs de l'Academie royale des Sciences. Avec figures en taille-douce* [Descriptions of the arts and trades, made or approved by the members of the royal academy of sciences. With figures in intaglio]. Paris. Großfol. The beginning was made in 1761, and since then, many individual

It inspired several more similar – and in part very valuable – works, not only in French, but also in other languages.[110] But how much is still lacking in all

descriptions of different handcrafts have been printed. Since 1771, the reformed priest Bertrand at Neufchatel has organized a reprint with many errors. This reprint seems to have grown to 17 volumes. – The German translation is called: *Schauplatz der Künste und Handwerke, oder vollständige Beschreibung derselben, verfertigt und gebilligt von der Akademie der Wissenschaften zu Paris*. From the year 1762 through 1805, 21 volumes in quarto.

110 93) Above all, the following: *Encyclopédie, ou Dictionnaire raisonné des sciences et des arts* [Encyclopedia, or reasoned dictionary of the sciences and the arts]. Paris and Neufchâtel. Folio. From the year 1751 through 1767, 17 volumes text and 11 volumes engravings. – *Supplement à l'Encyclopédie: par une societé des gens de lettres* [Supplement to the Enyclopedia, by a society of scholarly individuals]. Amsterdam 1778–1777. Folio. Four volumes text and one volume engravings. From the entire work, there are reprints from Geneva, Lucca, Florence, Yverdon, and Lyon. – *Tables alphabetiques des arts et metiers, contenus dans les douze volumes de planches de l'Encyclopédie* [Alphabetical tables of the arts and trades, contained in twelve volumes of plates]. Folio. – *Table analytique et raisonnée des matières contenues dans les 33 volumes in folio du Dictionnaire des sciences, des arts et des metiers et dans son supplement* [Analytical and reasoned table of the materials contained in the 33 folio volumes of the *Dictionary of the Sciences, Arts, and Trades* and in its supplement]. Paris 1780. Folio. Two volumes. – *Dictionnaire raisonné universel des arts et metiers* [Universal reasoned dictionary of the arts and trades]. New edition, edited by Abbé Joubers. Paris 1773. 9. Five volumes. – Encyclopédie, ou dictionnaire universel raisonné des connoissances humaines: mis en ordre par M. de Felie. Yverdon 1770–1776. 4. 24 volumes text, alongside six volumes of supplements and six volumes of engravings. – *Dizionario delle arti e de'mestieri, delle fabbriche, e delle manifatture* [Dictionary of the arts, trades, fabricks, and manufactures]. Compiled by Francesco Guiselini. Venice, 1768. 8. Was planned for 14 volumes. – Johann Gottfried Halle, *Werkstätte der heutigen Künste, oder die neue Kunsthistorie* [Workshops of the modern-day arts, or the new history of art]. Brandenburg und Leipzig 1761–1779. Six volumes in 4. – Peter Nathanael Sprengel. *Handwerke und Künste in Tabellen* [Handworks and arts in tables]. Continued by O. L. Hartwig. Berlin 1767–1795. Seventeen collections in 8. The first two collections were completely revised in 1779 and 1782 into a new edition, organied by Herr Hartwig. Both of the latter collections, however, have other authors. – J. G. H. von Justi, *Vollständige Abhandlung von den Manufakturen und Fabriken* [Complete treatise of manufactures and fabricks]. With corrections and comments by J. Beckmann. Two volumees. Berlin 1780. 8. A Dutch translation printed in Utrecht dates from the year 1783. – *Encyclopédie methodique, ou par ordre de matieres, par une societé de gens de lettres, de savans et d'artistes* [Methodological encyclopedia, or by order of materials]. By a society of scholarly individuals, intellectuals and artisans. Paris, in 4. Since the year 1782, Sixty-five deliveries. – *Volledige beschrijving van alle Konsten, Ambachten, Handwerken, Fabrieken, Trasieken* [Complete description of all arts, crafts, handcrafts, fabricks, trades]. Dordrecht 1788–1799. 8. Sixteen pieces. – J. G. Krünitz, *Ökonomisch-technologische Encyclopädie, oder allgemeines System der Staats-Stadt- Haus- und Land-Wirthschaft, und der Kunstgeschichte* [Economical-technological encyclopedia, or general system of the economics of land, town, house, and land, and the history of art]. 1773–1806. One hundred and two volumes in 8. Volumes one through seventy-five exist in a new edition. And from the entire work there is an excerpt in

of these efforts! It is true that the arts and handcrafts which the members of the Parisian Academy pursued for a long time, or at least had an opportunity to observe directly, could contain a precise depiction of what concerns the setting, organization, and systems of instruments and machines, as well as the working material, the preparation and composition of materials. But the esteemed academics had to write some treatises and essays based on reports that were sent in to them, whose correctness and precision could not always be guaranteed since they naturally could come from men who did not have the required expertise, nor the correct and practiced gift of observation. One could consequently suspect that among the descriptions of handcrafts and arts, various inaccuracies, prejudices, misapprehensions, and errors could have crept in. And this had to be all the more dangerous, as even the French noted, because the stamp of truth and reliability was imprinted upon them by the stamp of such a famous society.

Among the many books and writings which we possess on arts and handcrafts,[111] there are unfortunately not many wherein the subject would be described and depicted with complete clarity and precision, and still fewer which show the means and the paths of improving arts and handcrafts to simplify them and bring them closer to perfection. Naturally, the study of arts and handcrafts is of the greatest difficulty, and presupposes talents, insights, and disposition of mind, which only seldom unite in one and the same man. And furthermore, the cultivation of this subject demands a cost and time which not everyone can finance.

§53 That the Germans have long since, and particularly also in the *eighteenth* century, made themselves very useful, we know already from the prior discussion. One thanks them the most for a precise scientific instruction in this field. As proof, I need only to mention the names *Marperger*,

twenty-seven pieces. – *Deutsche Encyclopädie, oder allgemeines Realwörterbuch der Künste und Wissenschaften* [German encyclopedia, or general subject dictionary of the arts and sciences]. Frankfurt am Main. 1778–1805. Twenty-three volumes. Folio.

111 94) Among these number especially the following very valuable periodical works. C. W. J. Gatterer's *Technologisches Magazin* [Technological magazine]. From 1790, three volumes in octavo. – *Journal für Fabrik, Manufaktur, Handlung und Mode* [Journal for fabricks, manufacture, trade, and fashion]. Leipzig 1792–1806. Thirty volumes in 8. A small volume is published monthly. From the first volumes, a new edition has already been organized. – The *Repertory of Arts and Manufactures*. London 1794–1805. XXIII volumes in 9. – *Annales des arts et manufactures, ou memoires technologiques sur les découvertes modernes concernant les arts, les manufactures.* [Annals of arts and manufactures, or technological memoires on modern discoveries concerning the arts and the manufactures]. By R. Oreilly. Paris 1790. Several volumes in 8.

Kunkel, Zink, Schreber, von Justi, von Pfeifer, Bergins, Sprengel, Jacobsson, Jung, Beckmann, von Lamprecht, Krünitz, Flörke, Catterer, Böhmer, Schäfer, Wehrs, Rösig, Klaproth, von Born, Seezen, Pörner, Caucrin, Walther, Gädike, Nemnich, Lehmann, Langsdorf, Keserstein, Bautsch, Neuenhahn and *von Meidinger*. Among the Swedish people, these have distinguished themselves the most: *Polhem, Cederhielm, Fischerström, Elvius, Gadolin, Rinman* and *Nordwall*; of the English: *Lee, Kay, Sholl, Koops, Wedgwood, Arkwright, Cartwright, Barrington*, and *Bolton*; from the French: *Gobelin, Thevart, Vaucanson, Roland de la Platière, Bose, d'Antic, Demachy, Palissy, Hellot, Loysel, Macquer, du Hamel, von Reaumur, Chaptal, Bertholet, Felix Beaujour* and *Oreilly*.

The theory of handcrafts has gained much in recent times, above all through the efforts of *Hofmann, Weigel, Pott, Schlüter, Wiegleb, Erxleben, Achard, Cartheuser, Weber, Model, Succcow, Schreber, Scheele, Karsten, Gemelin, Gren, Pörner, von Dietrich, Werner, Gülich, Westrumb, Trhomsdorf, Scherer, von Crell, Langsdorf, Lampadius, Hermbstädt, Gehlen, Achard, le Sage, Macquer, Lavoisier, Bertholet, Chaptal, Fourcroy, Vauquelin*, and others, who have applied mathematics and the study of nature, particularly mathematics and chemistry, with advantage to the various technological works, and who made their scientific support ever firmer.

§54 The scientific treatment of the arts and handcrafts, however uncommonly useful it has been for the improvement of these trades, has also, from a different perspective, produced many abuses which hindered progress more than furthering it. To this matter belong, among others, the superfluous production of systems and hypotheses. Many attempts succeed on a small scale, and are unable to be executed on a large scale, or at least do not come with any particular advantage. Just as various machines and models look good, and when executed on a large scale lead to an uncertain or non-advantageous result. The talented Frenchman Reaumur invented the art of transforming melted iron into steel. One considered this art, among its kind, as a masterpiece, but was not in the position of practicing this art on a large scale to advantage. All those who made such an attempt according to his instruction have spent much, and have likely been ruined in the process. The cause of this should be obvious. Reaumur practiced the preparation of steel in his small laboratory, rather than trying it in a large ironworks. For only in the actual workshops can one learn about all the conditions of the operations. There one first sees various things from another perspective, there one can observe and compare, change, and improve at various times and under various conditions, there one can properly succeed in making adjustments, mixtures, separations, etc., if they are ever to be achieved.

Many scholars and other men who wanted to acquire a precise knowledge of handcrafts, arts, manufacture, and fabrick works, have often been led astray in that they have relied too much on the workers. They often came to great harm thereby, and they often brought themselves and others into their misfortune. A complete knowledge of the execution of handcrafts and arts, alongside all of the occupations and circumstances that occur alongside, can only be the fruit of long experience, of a tireless patience, of an iron industriousness and a great spirit of observation. And up until now, very few have acquired these qualities.

With the introduction of new inventions or other improvements and changes, one has the most reason to be on guard. Here, one can never dispense with the most thorough consideration. Just as many a new device which began with eagerness and the most glorious hopes has in recent years sunk back down to nothing! Many otherwise intelligent men believed to have achieved a great gain, that they saved force and effort with their works or undertakings, and afterwards discovered to their horror the much greater loss in time, which destroyed all of their hopes again. Meanwhile, effort and time were both spared, but the wares came out all the worse. Among the many examples which could be mentioned here from past times, I give only a few. A director of the famous mirror fabrick at St. Gobain in France arranged for a new smelting oven, from which one should do *four* pours each time, instead of *three*, as usual. The idea worked in its execution in so far as *four* pours were really made from a single oven loading, and thus each week one received twenty-four pieces of glass more. But this advantage was only illusory, for by closer investigation it turned out that the pouring, smelting, and purifying in the new device took twice as much time as before. One required namely seventy to seventy-two hours of time, where one otherwise needed only twenty-four. Subsequently, one could only fill the oven twice per week. Apart from that, the glasses were completely brown-green, an unpleasant color, poorly and shot through with little stones, and almost none of them could endure evening and polishing. – In a few brandy distilleries, one changed in the last years of the eighteenth century the ovens, in order to have them save wood, based on a certain suggestion. One soon returned these ovens, however, because the distillation now went more slowly and was a greater disadvantage to the distillers than the advantage of saving wood had contributed. – In Münzen one has a steel mill which achieved stamping very quickly, but the stamp was very weak and the stamped coins were concave and not entirely circular. – Such a performance or construction produces a more complete ware than in any other way, but it demands too much finesse, too much foresight, for it to be usefully applied. Beautiful, expensive mirrors of the most various kinds are often achieved only after many vain pours, and for that reason one does them best at medium size in fabricks.

One has also, over time, and particularly in the eighteenth century, become much more talented and ingenious, but not always much smarter, because one has often not thought about what the mechanic calls *economical effect*. Also in all of the arts and handcrafts, one should not ignore this effect.

With much success, a pair of men, Gädike[112] and Nemnich[113] made the attempt at putting together an exact inventory of fabricks and manufactures and the wares produced in them. Such inventories are certainly of use and can be even more so if one continuously seeks to supplement and improve them. The various branches of technology have naturally become very large, extensive, and beautiful, but they can still keep increasing in size, usefulness, and beauty!

112 95) Johann Christian Gädike. *Fabriken- und Manufakturen- Adreß- Lexikon von Teuschland.* [Factories and Manufactures address-lexicon of Germany]. Two parts. Weimar 1799. 8. New edition.

113 96) Philipp Andreas Nemnich. *Waaren-Lexicon in zwölf Sprachen* [Lexicon of wares in twelve languages]. Part one. Hamburg. 1797. 8. Part two. Hamburg. 1801. 8. Part three. Hamburg. 1802. 8.

Bibliography

Adelung, Johann Christoph. *Historisch-Critisches Wörterbuch der hochdeutschen Mundart* (Historical-critical dictionary of the German language). B. Ph. Bauer, 1811. https://lexika.digitale-sammlungen.de/adelung/online/angebot.

d'Alembert, Jean-Baptiste le Rond. "Preliminary Discourse." The Encyclopedia of Diderot & d'Alembert Collaborative Translation Project. Translated by Richard N. Schwab and Walter E. Rex. Ann Arbor: Michigan Publishing, University of Michigan Library, 2009. http://hdl.handle.net/2027/spo.did2222.0001.083 (accessed August 26, 2023). Originally published as "Discours Préliminaire," Encyclopédie ou Dictionnaire raisonné des sciences, des arts et des métiers, 1:i–xlv. Paris, 1751.

Angier, Tom. Technē *in Aristotle's Ethics: Crafting the Moral Life*. Continuum, 2012.

Ausonius. Translated by Hugh G. Evelyn White. Vol. 1. William Heinemann, 1919.

Bacon, Francis. *The Advancement of Learning*, edited by William Aldis Wright. Fifth edition. Clarendon Press, 1920.

Bacon, Francis. *The New Organon*, edited by Lisa Jardine and Michael Silverthorne. Cambridge University Press, 2000.

Bailey, Nathan. *Dictionary or Wörterbuch*, 9th edition, vol. 2. Friedrich Frommann, 1797.

Barney, Rachel. "*Technê* As a Model for Virtue in Plato." In *Productive Knowledge in Ancient Philosophy*, edited by Thomas Kjeller Johansen. Cambridge University Press, 2021.

Beckert, Manfred. *Johann Beckmann*. Teubner, 1983.

Beckmann, Johann. *Anleitung zur Technologie oder zur Kenntniß der Handwerke, Fabriken und Manufacturen, vornehmlich derer, die mit der Landwirthschaft, Polizey und Cameralwissenschaft in nächster Verbindung stehn* [Guide to technology or to the knowledge of handcrafts, fabricks, and manufactures, primarily those which are most closely connected with agriculture, science of governance, and cameralist science]. First edition. Widow Vandenhoeck, 1777.

Beckmann, Johann. *Beyträge zur Geschichte der Erfindungen* [Contributions to the history of inventions]. Paul Gotthelf Kummer, 1780.

Beckmann, Johann. "Entwurf einer algemeinen (sic) Technologie." In *Vorrath kleiner Anmerkungen über mancherley gelehrte Gegenstände* [Collection of small observations concerning various scholarly topics]. Vol. 3. Johann Friedrich Röwer, 1806.

Beckmann, Johann. *La technologie générale*. Edited by Guillaume Carnino, Liliane Hilaire-Pérez, and Jochen Hoock. Translated by Joost Mertens. Introduction by Guillaume Carnino and Liliane Hilaire-Pérez. Presses Universitaires, 2017.

Beckmann, Johann. *Physikalisch-ökonomische Bibliothek*. Vol. 3. Widow Vandenhoeck, 1772.

Beckmann, Johann. *Physikalisch-ökonomische Bibliothek*. Vol. 4. Widow Vandenhoeck, 1773.

Beckmann, Johann. *Physikalisch-ökonomische Bibliothek*. Vol. 5. Widow Vandenhoeck, 1774.

Bensen, Daniel Heinrich. *Versuch eines systematischen Grundrisses der reinen und angewandten Staatslehre für Kameralisten* [Attempt at a systematic outline of pure and applied political theory]. First edition. Johann Jakob Palm, 1798.

Bertrand, Jean Élie. "Préface." In *Descriptions des arts et métiers* [Descriptions of arts and trades]. New edition. Volume 1. Société Typographique, 1771.

Böning, Holger. "Reimarus, Johann Albrecht Heinrich, 1729–1814." In *The Bloomsbury Dictionary of Eighteenth-century German Philosophers*. Edited by Heiner F. Klemme and Manfred Kuehn. Bloomsbury, 2016.

Breazeale, Daniel. "Note on translation." In Johann Gottlieb Fichte, *Early Philosophical Writing*. Cornell University Press, 1988.

Brewer, Daniel. "The *Encylopédie*: Innovation and Legacy." In *New Essays on Diderot*. Edited by James Fowler. Cambridge: Cambridge University Press, 2011.

Brisson, Luc. "Myth and Knowledge." Translated by Elizabeth Rawlings and Jeannine Pucci. In *Greek Thought: a Guide to Classical Knowledge*. Edited by Jacques Brunschwig and Geoffffrey E. R. Lloyd with Pierre Pellegrin. Translated by Catherine Porter. Harvard University Press, 2000.

Brunschwig, Jacques. "Epistemology." Translated by Catherine Porter and Dominique Jouhaud. In *Greek Thought: a Guide to Classical Knowledge*. Edited by Jacques Brunschwig and Geoffffrey E. R. Lloyd with Pierre Pellegrin. Translated by Catherine Porter. Harvard University Press, 2000.

De Bruyn, Frans. "Introduction." In *The Cambridge Companion to Eighteenth-century Thought*. Edited by Frans de Bruyn. Cambridge University Press, 2021.

Burford, Alison. *Craftsmen in Greek and Roman Society*. Cornell University Press, 1972.

Cadden, Joan. "The Organization of Knowledge: Disciplines and Practices." In *The Cambridge History of Science*, vol. 2. Medieval Science. Edited by David C. Lindberg and Michael H. Shank. Cambridge University Press, 2013.

Candel, Danielle and John Humbley, "L'émergence d'une conscience terminologique: regard métalexicographique porté sur l'Entwurf der algemeinen Technologie de Johann Beckmann (1806)" [The emergence of a terminological conscience: a metalexicographic look at the draft of a general technology by Johann Beckmann]. In *La technologie Générale*. Edited by Guillaume Carnino, Liliane Hilaire-Pérez, and Jochen Hoock. Translated by Joost Mertens. Introduction by Guillaume Carnino and Liliane Hilaire-Pérez. Presses Universitaires, 2017.

Coffin, Sarah D. "Servers, Sugar." In *The Oxford Companion to Sugar and Sweets*. Edited by Darra Goldstein. Oxford University Press, 2015.

Coles, Elisha. *An English Dictionary*. 1st edition. Samuel Crouch, 1676.

Cuomo, Serafina. *Technology and Culture in Greek and Roman Antiquity.* Cambridge University Press, 2007.

Diderot, Denis (attributed). "Art." The Encyclopedia of Diderot & d'Alembert Collaborative Translation Project. Translated by Nelly S. Hoyt and Thomas Cassirer. Michigan Publishing, University of Michigan Library, 2003. http://hdl.handle.net/2027/spo.did2222.0000.139 (accessed August 16, 2023). Originally published as "Art," Encyclopédie ou Dictionnaire raisonné des sciences, des arts et des métiers, 1:713–717 (Paris, 1751).

Eberhard, Johann August. *Synonymisches Handwörterbuch Der Deutschen Sprache* [Synonymous pocket dictionary of the German language]. Thirteenth edition. Grieben, 1882.

Ebers, John. *The New and Complete Dictionary of the German and English Languages.* 3 vols. Breitkopf and Haertel, 1796–1799.

Epstein, S. R. and Maarten Prak. *Guilds, Innovation and the European Economy, 1400–1800.* Cambridge University Press, 2008.

Farr, James R. *Artisans in Europe, 1300–1914.* Cambridge University Press, 2001.

Fox, Celina. *The Arts of Industry in the Age of Enlightenment.* Yale University Press, 2009.

Frambach, Hans. "The Decline of Cameralism in Germany at the Turn of the Nineteenth Century." In *Cameralism in Practice. State Administration and Economy in Early Modern Europe,* edited by Marten Seppel and Keith Tribe. The Boydell Press, 2017.

Franzel, Sean, Ilinca Iurascu, and Petra McGillen, "Media Inventories of the Nineteenth Century: A Report from Two Workshops." *Goethe Yearbook* XXVIII (2021). 285–296.

Friedrich the Great, "Anti-Machiavell." In *The Works of Nicholas Machiavel* [sic], vol. 1, translated by Ellis Farenworth. Thomas Davies, 1762.

Frison, Guido. "Linnaeus, Beckmann, Marx and the Foundations of Technology. Between Natural and Social Sciences: a Hypothesis of an Ideal Type." *History and Technology* 10 (1993): 139–73.

Fröhner, Annette. "'Wenige, welche lesen mögen, werden nicht mit Nutzen und Vergnügen dieses Buch brauchen können' [Few readers will find this book devoid of usefulness and pleasure]. Johann Beckmann und Johann Georg Krünitz: Reflexionen und Einflüsse." In *Johann Beckmann (1739–1811).* Edited by Günter Bayerl and Jürgen Beckmann. Waxmann, 1999.

de Gandt, François. "Technology." In *Greek Thought: a Guide to Classical Knowledge.* Edited by Jacques Brunschwig and Geofffrey E. R. Lloyd, with Pierre Pellegrin. Translated by Selina Stewart and Jeannine Pucci. Harvard University Press, 2000.

Garrioch, David. "Artisan Mobility and Innovation in Pre-industrial Europe." In *The Republic of Skill. Artisan Mobility, Innovation, and the Circulation of Knowledge in Premodern Europe.* Edited by David Garrioch. Brill, 2022.

Genovesi, Abbé. *Delle lezioni di commercio o sia d'economia civile.* Federico Agnelli, 1768.

von Goethe, Johann Wolfgang. "Entdeckungen eines trefflichen Vorarbeiters" [Discovery of an excellent predecessor]. In *Sämtliche Werke*, vol. 12, *Zur Naturwissenschaft überhaupt, besonders zur Morphologie* [On the study of nature in general, particularly on morphology]. Edited by Hans J. Becker, Gerhard H. Müller, John Neubauer, and Peter Schmidt. 80–86. Hanser Verlag, 1989.

Grimm, Jacob and Wilhelm. *Deutsches Wörterbuch*. http://dwb.uni-trier.de/de/. 1854–1971.

Harris, John. *Lexicon technicum: or, an Universal English Dictionary of Arts and Sciences*, vol. 1. Dan Brown, 1704.

Hellmüth, Eckhart. "Der Staat des 18. Jahrhunderts." In *Aufklärung*, vol. 9, part 1. Edited by Günter Birtsch. Meiner, 1996.

Hermann (sic), Benedikt Franz. "Introduction." In *Ueber die Einführung des Studiums der Technologie, oder über die Lehre von Handwerken, Künsten, Manufakturen und Fabriken*. Bernardische Buchhandlung, 1781.

Hilaire-Pérez, Liliane. "Technology as a Public Culture in the Eighteenth Century: The Artisans' Legacy." *History of Science* 45 (2007): 135–153.

Hill, John. *A History of the Materia Medica*. T. Longman, 1751.

H. J. P. "Petit-Maitres." In *The Monthly Mirror: Reflecting Man and Manners* 17 (May 1804): 296–7.

Hoock, Jochen. "Johann Beckmann et la place du discours technologique dans la pensée économique de la deuxième moitié du xviiie siècle." In *La technologie générale*. Edited by Guillaume Carnino, Liliane Hilaire-Pérez, and Jochen Hoock. Translated by Joost Mertens. Introduction by Guillaume Carnino and Liliane Hilaire-Pérez. Presses Universitaires, 2017.

Hugh of St. Victor. *The Didascalicon of Hugh of St. Victor*. Translated, with introduction, by Jerome Taylor. Columbia University Press, 1961.

Hugon, Edouard. *Cosmology*. Translated by Romero Carrasquillo. editions scholasticae, 2013.

Jamison, Andrew. "Technology's Theorists: Conceptions of Innovation in Relation to Science and Technology Policy." *Technology and Culture* 30, no. 3 (July 1989): 505–533.

Johansen, Thomas Kjeller. *Productive Knowledge in Ancient Philosophy. The Concept of* Techné. Cambridge University Press, 2021.

Klippel, Diethelm. "Reasonable Aims of Civil Society: Concerns of the State in German Political Theory in the Eighteenth and Early Nineteenth Centuries." In *Rethinking Leviathan*. Edited by John Brewer and Echkhart Hellmüth. Oxford University Press, 1999.

Knemeyer, Franz-Ludwig. "Polizei." Translated by Keith Tribe in *Economy and Society*, vol. 9 (1980): 172–206.

Krünitz, Johann Georg. *Oekonomische Encyklopädie*. 242 volumes. 1773–1858. https://www.kruenitz1.uni-trier.de/.

Van der Laan, J. M. *Narratives of Technology*. Macmillan, 2016.

Lamprecht, Georg. *Lehrbuch der Technologie oder Anleitung zur Kenntniß der Handwerke, Fabriken und Manufacturen* [Instruction book of technology or guide to the knowledge of handcrafts, fabricks, and manufactures]. Hemmerdesche Buchhandlung, 1787.

Landreth, Sara. 2021. "Science in the Long Eighteenth Century." In *The Cambridge Companion to Eighteenth-century Thought*. Edited by Frans de Bruyn, Cambridge University Press, 2021.

Leibniz, Gottfried. Leibniz. *Unvorgreifliche Gedanken betreffend die Ausübung und Verbesserung der Deutschen Sprache* [Unpresumptuous thoughts having to do with the use and improvement of the German language]. Christian Georg Ackermann, 1831.

Leitch, Vincent B. "Postmodern Theory of Technology: Agendas." *symplokē*, vol. 12, no. 1 and 2 (2004): 209–215.

Liddell, George Henry and Robert Scott. *A Greek-English Lexicon*. http://www.perseus.tufts.edu/hopper/ (author note: the website does not specify which edition they put online).

Long, Pamela O. *Artisan/Practitioners and the Rise of the New Sciences, 1400–1600*. Oregon State University Press, 2011.

Long, Pamela O. *Technology, Society, and Culture in Late Medieval and Renaissance Europe, 1300–1600*. SHOT Historical Perspectives on Technology. American Historical Association, 2000.

Loveland, Jeff. "The *Dictionnaire universel de commerce (1723–1730)*." In *Specialized Dictionaries and Encyclopedias, 1650–1800*. Edited by Jeff Loveland and Stéphane Schmitt. Liverpool University Press, 2024.

Maar, J. H. "Johann Beckmann (1739–1811) and Modern Chemical Technology." *Substantia*, vol. 7.1 (2002): 79–92.

Marx, Karl. *Capital*, vol. 1. Translated by Eden Paul. Dent. International Publishers, 1996.

Marx, Leo. "Technology. The Emergence of a Hazardous Concept." *Technology and Culture*, vol. 51. July (2020): 561–577.

Meyer, Torsten. *Natur, Technik und Wirtschaftswachstum im 18. Jahrhundert. Risikoperzeptionen und Sicherheitsversprechen*. Münster, 1999.

Mitcham, Carl. *Thinking through Technology. The Path Between Engineering and Philosophy*. University of Chicago Press, 1994.

Müller, Hans Peter. "Johann Beckmann und Carl von Linné – Anmerkungen zu einer prägenden Begegnung" [Notes on a formative encounter]. In *Johann Beckmann (1739–1811)*, Edited by Günter Bayerl and Jürgen Beckmann. Waxmann, 1999.

Multhauf, Robert P. "The Historiography of Technology: Some Observations on the State of the History of Technology." *Technology and Culture* 15, no. 1 (January 1974): 1–12.

De Munck, Bert. "Shared Skills and Technologies of Community Formation. Artisanal Epistemologies, Secrecy, and Governmentality in Long-Term Context." In *The Republic of Skill. Artisan Mobility, Innovation, and the Circulation of Knowledge in Premodern Europe*. Edited by David Garrioch. Brill, 2022.

Ogilvie, Sheilagh. *The European Giulds. An Economic Analysis*. Princeton University Press, 2019.

Ortega y Gassett, José. *Toward a Philosophy of History*. University of Illinois Press, 2002.

Ovitt, George. *The Restoration of Perfection. Labor and Technology in Medieval Culture*. Rutgers University Press. 1987.

Ovitt, George. "Technology and Science." In *The Cambridge History of Science*, vol. 2, Medieval Science. Edited by David C. Lindberg and Michael H. Shank. Cambridge University Press, 2013.

Peters, John Durham. "You Mean My Whole Fallacy Is Wrong." In *Representations* 140 (Fall 2017): 10–26.

Picon, Antoine. "Towards a History of Technological Thought." In *Technological Change: Methods and Themes in the History of Technology*. Edited by Robert Fox. Routledge, 1996.

Plato. *The Republic of Plato*. Translated by Allan Bloom. Basic Books, 1991.

Poppe, Johann Heinrich Moritz. *Ausführliche Volks-Gewerbs-Lehre oder allgemeine und besdondere Technologie* [Complete doctrine of popular industry or general and specific technology]. Hoffmann, 1839.

Poppe, Johann Heinrich Moritz. *Geschichte der Technologie seit der Wiederherstellung der Wissenschaften bis an das Ende des achtzehnten Jahrhunderts* [History of technology since the restoration of the sciences through the end of the eighteenth century], vol. 1. Johann Friedrich Röwer, 1807.

Pseudo-Geber, *The Summa perfectionis of Pseudo-Geber: A Critical Edition, Translation and Study*. Edited and translated by William R. Newman. Brill, 1991.

Rabenhorst, C. T. *New Pocket-Dictionary of German and English Languages*. C. T. Rabenhorst, 1800.

Roche, Daniel. "Encyclopedias and the Diffusion of Knowledge." In *The Cambridge History of Eighteenth-century Political Thought*. Edited by Mark Goldie and Robert Wokler. Cambridge University Press, 2006.

Roochnik, David. *Of Art and Wisdom. Plato's Understanding of Techne*. Penn State University Press, 1996.

Rossi, Paolo. *Philosophy, Technology, and the Arts in the Early Modern Era*. Torchbook, 1970.

Schatzberg, Eric. "'Technik' comes to America: Changing Meanings of Technology Before 1930." In *Technology and Culture* 47, no. 3 (July 2006): 486–512.

Schatzberg, Eric. *Technology. Critical History of a Concept*. University of Chicago Press, 2018.

Schmitt, Stéphane. 2024. "Jacques-Christophe Valmont de Bomare's *Dictionnaire raisonné universel d'histoire naturelle* (1764)." In *Specialised Dictionaries and Encyclopedias, 1650–1800*. Edited by Jeff Loveland and Stéphane Schmitt. Liverpool University Press, 2024.

Schultes, Joseph August. *Ausflüge nach dem Schneeberge in Unterösterreich* [Excursions to Schneeberg in Lower Austria]. J. V. Degen, 1802.

Seibicke, Wilfried. "Von Christian Wolff zu Johann Beckmann." In *Mehrsprachigkeit in der deutschen Aufklärung* [Multilingualism in the German Enlightenment]. Edited by Dieter Kempel. Meiner, 1985, 52–61.

Seibicke, Wilfried. *Technik. Versuch einer Geschichte der Wortfamilie* [Technics. Attempt at a history of the word family]. VDI-Verlag, 1968.

Sennett, Richard. *The Craftsman*. Yale University Press, 2008.

Seppel, Marten. "Cameralism in Practice." Introduction to *Cameralism in Practice. State Administration and Economy in Early Modern Europe*. Edited by Marten Seppel and Keith Tribe. Boydell and Brewer, 2017.

Seubert, Rolf. "Handwerkbildung als Beitrag zum Aufbau einer Nationalindustrie. Johann Heinrich Moritz Poppe und der Zusammenhang von Technologieentwicklung und beruflicher Bildung" [Handcraft education as contribution to the building of a national industry]. In *Johann Beckmann (1739–1811)*. Edited by Günter Bayerl und Jürgen Beckmann. Waxmann, 1999.

Strelin, Georg Gottfried. *Versuch einer Geschichte und Literatur der Staatswissenschaft*. Erlangen: J. J. Palm und E. Enke, 1827.

Taws, Richard and Genevieve Warwick. "After Prometheus: Art and Technology in Early Modern Europe." In *Art and Technology in Early Modern Europe*, edited by Richard Taws and Genevieve Warwick, Wiley-Blackwell, 2016.

Thomasius, Christian. *Einleitung zu der Vernunfft-Lehre*. Christoph Salfeld, 1691.

Tribe, Keith. "Cameralism and the Sciences of the State." In *The Cambridge History of Eighteenth-Century Political Thought*. Edited by Mark Goldie and Robert Wokler. Cambridge University Press, 2006.

Tribe, Keith. "Cameralism and the Science of Government." In *The Journal of Modern History* 56.2 (June 1984): 263–284.

Tribe, Keith. *The Economy of the Word. Language, History, and Economics*. Oxford University Press, 2015.

Troitzsch, Ulrich. *Ansätze technologischen Denkens bei den Kameralisten des 17. Und 18. Jahrhunderts* [Approaches to technological thinking by the cameralists of the 17th and 18th centuries]. Duncker und Humboldt, 1966.

Troitzsch, Ulrich. "Manufakturen in Deutschland und ihre theoretische Behandlung in der kameralistischen Literatur. Ansätze zu einem Vergleich" [Manufactures in Germany and their theoretical treatment in the cameralist literature. Approaches to a comparison]. In *Wirtschaftskräfte und Wirtschaftswege. Festschrift für Hermann Kellenbenz*. Edited by J. Schneider. Klett Cotta, 1978.

Wakefield, Andre. *The Disordered Police State. German Cameralism as Science and Practice*. The University of Chicago Press, 2009.

Weber, Wolfhard. "Die 'Anleitung zur Technologie' von Johann Beckmann." In *Technologie zwischen Fortschritt und Tradition*. Edited by Hans-Peter Müller and Ulrich Troitzsch. Peter Lang, 1992.

Weber, Wolfhard. "Technik zwischen Wirtschaft und Handwerk. Die Technologie des 18. Jahrhunderts als Lenkungswissenschaft des spätabsolutischen Staates" [Technics between economy and handcraft. The technology of the 18th century as guiding science of the late absolutist state]. In *Wirtschaft, Technik und Geschichte: Beiträge zur Erforschung der Kulturbeziehungen in Deutschland und Osteuropa*. Edited by Albrecht Timm, Volker Schmidtchen, and Eckhard Jäger. Camen, 1980.

Whitney, Elspeth. *Medieval Science and Technology*. Greenwood Press, 2004.

Zedler, Johann Heinrich, *Grosses vollständiges Universal-Lexikon aller Wissenschaften und Künste* [Great complete universal-lexicon of all sciences and arts], vol. 42. Johann Heinrich Zedler, 1744.

Index

Printed in the United States
by Baker & Taylor Publisher Services